SPECIAL
RELATIVITY

World Scientific Lecture Notes in Physics Vol. 33

SPECIAL RELATIVITY

Ulrich E. Schröder

Johann Wolfgang Goethe-Universität
Frankfurt am Main, Germany

World Scientific

Singapore • New Jersey • London • Hong Kong

Published by

World Scientific Publishing Co. Pte. Ltd.,
P O Box 128, Farrer Road, Singapore 9128
USA office: 687 Hartwell Street, Teaneck, NJ 07666
UK office: 73 Lynton Mead, Totteridge, London N20 8DH

Library of Congress Cataloging-in-Publication Data

Schröder, Ulrich E.
　　Special relativity/by U. E. Schröder.
　　p. cm.
　　ISBN 9810200684.　--　ISBN 981020132X (pbk.)

　　1. Special relativity (Physics)　　I. Title.
　　QC173.65.S35　　　1990
　　530.1'1 -- dc20　　　　　　　　　　　　　　　　　　　89-48134
　　　　　　　　　　　　　　　　　　　　　　　　　　　　　CIP

Printed in Singapore by JBW Printers and Binders Pte. Ltd.

PREFACE

This book originated from lectures which I gave repeatedly at the University of Frankfurt am Main. The lively interest of the students testified for the need of such a text. Because of the multitude of topics which make up the curriculum, special relativity gets short shrift. On the other hand, monographs in this area are very voluminous and serve better as reference works. In contrast, it should be possible to read through this book on hand in a relatively short time. It contains the material which was treated in a semester of a two-hour-per-week course. However, the presentation is more detailed than it would be in a set of lecture course notes, so that it may serve for private study too.

Since the German version of the book had a very positive reception, it soon became necessary to prepare a new and enlarged edition. The present English text is based on the second German edition, which was published in 1987.

The book provides a short course in special relativity, intended mainly for senior undergraduates or beginning graduate students in physics, mathematics, or related subjects such as astrophysics. It is assumed that the reader already has some basic familiarity with standard course materials. Thus, for example, the famous Michelson-Morley experiment is not once again described in detail.

The advanced reader will soon notice that the emphasis in this work is different from that of most presentations, especially from the so well established older texts. Via the path I follow, one arrives at the essential statements of the theory by a more direct approach. Thus, access to the special theory of relativity is rendered exactly in the spirit of A. Einstein's comment, given on page ten.

After some introductory remarks on the essence of relativity theory the book begins in Chapter 1 with a brief historical outline of the main contributions to its development by H.A.Lorentz, H.Poincaré, A. Einstein, and H. Minkowski. Also the reception of Einstein's crucial

work of 1905 is commented on, as well as the role of the Michelson-
Morley experiment in Einstein's reasoning. The Galilean relativity
principle and its limits are discussed in Chapter 2. For the establi-
shing of the Lorentz transformation the principle of relativity and
the homogeneity and isotroᵖⱼ of inertial frames suffice. In consequence of Einstein's addition law for velocities, which follows from
the Lorentz transformation, a certain universal velocity cannot be
exceeded. Only after having recognized the existence of this finite
limiting velocity, is the latter identified with the speed of light.
The kinematical consequences of the Lorentz transformation and the
often discussed "paradoxes" are given a detailed treatment by way
of examples in Section 2.5. Appropriate to the content of special
relativity is the covariant formulation of physical laws in terms
of relations between tensorial entities. Chapter 3 serves as an introduction to tensor calculus. The concept of a tensor field is introduced
without restriction to linear transformations, and hence this chapter
will prepare the student for the more general mathematical concepts
used in general relativity. Concerning the notation it should be mentioned that three-dimensional vectors are written with an arrow on
top of the character, whereas the components of four-vectors and four-
tensors respectively, are denoted by greek indices. When studying
Chapter 4 the reader will become familiar with the covariant formulation of relativity theory in Minkowski space. In this chapter the
different pieces of the full Lorentz group and also the geometrical
representation of the Lorentz transformation are discussed. Chapter
5 contains the relativistic formulation of classical mechanics. Special
attention is given to the equivalence of mass and energy, illuminating
this fundamental result by several examples. The relativistic invariant
Lagrangian allows for a convenient derivation of the conservation
laws. Energy-momentum conservation is exemplified by a variety of
different processes, such as decay, creation, and scattering of
particles. The attempt to formulate the dynamics of a system
of interacting relativistic particles leads quite naturally to the
concept of a physical field as the convenient means for a causal
propagation of physical effects, i.e. interactions. As the most promi-

nent example of a relativistic field theory, electrodynamics in its
covariant formulation is treated in Chapter 6. A brief derivation
of the fundamental Noether theorem is given, which shows how the conservation laws follow from the underlying symmetry transformations.
In our case the symmetrical energy–momentum tensor of the electromagnetic field may be derived via Noether's theorem, thus avoiding the
ad hoc and often used symmetrization procedure. In the subsequent
Chapter 7 the fundamentals of relativistic hydrodynamics are presented,
which are important for various topical applications in nuclear physics
as well as in astrophysics. Finally, the last chapter contains a discussion regarding the limits of validity of special relativity theory,
and it ends with a glimpse of the essential features of Einstein's
gravitation theory, i.e. the theory of general relativity. More about
the coverage of this book can be gleaned from the Table of Contents.

We quote a number of bibliographical data so as to guide the
reader to the specialized literature which will offer most profitable
and interesting further reading. After studying this text, he should
be able to better understand the materials presented in journals.
The problems given in the appendix are meant to stimulate further
thought. At the end of the book the reader also finds a tabulation
of more recent experiments testing the theory of special relativity.

I want to thank all readers of the German edition who called
my attention to possible improvements. Special thanks are due to
Professor Herbert Pfister (Tübingen), whose detailed comments proved
most helpful. I also take the opportunity to thank my esteemed teacher,
Professor Friedrich Hund (Göttingen), for his valuable inspirations.

I further want to thank Mrs. G. Boffo for the preparation of
the manuscript for the printer, including the figures and the formulae,
and Mrs. E. Martens for typing the footnotes.

Finally, I thank both my daughters, Ullinca and **Tine**, for their
help rendered, especially for drawing the vignettes at the chapter
endings.

Frankfurt/Main, September 1989   U.E.Schröder

CONTENTS

INTRODUCTION

In this book we study the physical foundations of our contemporary view of space and time. We also discuss consequences of the established spatial-temporal symmetry transformations. These symmetries, expressed via the Lorentz- or Poincaré group, are universal i.e., they apply to all physical systems and processes. Hence, it would appear necessary to study all branches of physics in regard to their behaviour relative to the Lorentz group. In this book we will focus on the following special areas of classical physics: relativistic mechanics of a mass point, electrodynamics as an example of a relativistic field theory, and finally, the basic equations of relativistic fluid dynamics.

The term "theory of relativity" is not a fortunate one, inasmuch as it paraphrases the essence of the theory in a rather negative manner, thereby giving rise to several misunderstandings. The terminology originated from the fundamental "relativity principle", formulated at the beginning of this century by Poincaré and Einstein. This principle became the foundation for the development of new ideas about space and time. It made it possible to refute the notion of an absolute non-moving "Ether", a concept that has been introduced earlier, in analogy to the theory of the elastic phenomena, in order to serve as a "medium" in which optical and electromagnetic phenomena take place. At that time, one attempted to explain all phenomena by some mechanical theory and in this framework one concluded that it should be possible to establish motion relative to an ether at absolute rest. This conception contradicted the fact (known already to Galileo (1564-1642) and Huygens (1629-1695)) that, within the framework of mechanics, only relative velocities can be measured between bodies in uniform motion. (This statement is the Galilean principle of relativity for mechanics.) The hypothesized existence of an ether at rest was eventually refuted by compelling experimental findings. It turned out that even very sensitive experiments related to electromagnetic processes (such as the Michelson-Morley experiment in 1887) could not indicate the presence of an absolute frame of reference at rest. In electromagnetism, just as in mechanics, one can observe only relative motion. This insight is formalized in the "relativity principle", annunciated in the writings of Poincaré, and formulated more consistently and deeply by Einstein as an universal law. Joining it with the additional "principle of constant light velocity", in his 1905 work Einstein developed these elements into the foundations of the theory of special relativity.

However, the essence of relativity theory is not so much the "relativization" of concepts such as space and time, but rather the insight that the laws of nature are independent of the choice of one or another frame of reference provided these are in uniform relative motion to each other. The decisive statement of the theory is that natural phenomena are invariant against the change of a frame of reference, provided this change is in accord with the transformations of the Lorentz group. This maxim clarifies to what extent it is possible to make absolute physical statements (i.e. statements independent from the frame of reference). Invariance with respect to the Lorentz group implies certain structural features of physical laws. Thus, the relativity principle may be looked upon as an organizing ingredient of nature's laws.

Chapter 1

HISTORY OF THE DEVELOPMENT OF
RELATIVITY THEORY

The history of development of special relativity theory is both interesting and instructive. We restrict ourselves to a brief review of the three crucial contributions of Hendrik Anton Lorentz (1853–1928), Henry Poincaré (1854–1912) and Albert Einstein (1879–1955); the breakthrough was achieved by these scientists in 1904–1905.[1] Even though the formal content of the quoted papers is rather similar, their motivations and the viewpoints that underly them are considerably different.

The concept of· an absolute ether at rest became suspect when Michelson and Morley could not succeed in determining the motion of the Earth relative to the ether at rest, despite the fact that the

---

[1] H.A. Lorentz, Proc.Acad.Sc. Amsterdam 6,809(1904);
H. Poincaré, Comp.Rend. 140, 1504 (1905) and Rendiconti Circolo
Mat. Palermo 21, 129 (1906);
A. Einstein, Ann. Physik 17, 891 (1905).

The papers of Lorentz and Einstein are reprinted in the volume entitled "H.A. Lorentz, A. Einstein, H. Minkowski, and H. Weyl: The Principle of Relativity - a Collection of Original Memoirs", trans. W. Perrett and G.B. Jeffery (Methuen & Co., London, 1923; paperback reprint, Dover Publications, 1958). An English translation (with commentary) of essential parts of the Rendiconti work of Poincaré may be found in the articles of H.M. Schwarz, Am. J. Phys. 39, 1287 (1971), 40, 862, and 1282 (1972).

sensitivity of their experiment sufficed to observe the effects predicted by the then current theory. In order to save the ether concept, G.F. Fitzgerald and, independently, Lorentz suggested the hypothesis that material bodies contract in the direction of their motion by a factor $\sqrt{1-v^2/c^2}$. (This is called the Lorentz-Fitzgerald contraction.) This assumption allowed for an interpretation of the null-result of the Michelson-Morley experiment.[1] But now Lorentz and also Poincaré had to face the problem: how could one explain the contraction in terms of a model of matter. We should note that contraction of distances occurs also in Einsteins framework. However, this does not require explanation via some model, since, according to the relativity principle, it arises already as a consequence of the observer's point of reference. In contrast, Lorentz provided an explanation for the contraction by an electromagnetic model of matter. This electron-theory of matter was expounded in the research contribution of 1904. This paper contains the Lorentz tranformation, and presents also the transformation rules for the electromagnetic field strengths $\vec{E}$ and $\vec{B}$. Lorentz introduced along with absolute time a so-called "local time", but he looked at this merely as a mathematical trick. He did not try to give local time some experimental meaning. The physical content of the transformation formulae was obscure. Lorentz insisted on an ether at rest even in 1910, and eventually gave up the notion of absolute simultaneity of events only with reluctance.

In his talk delivered at the International Congress of Arts and Sciences (St. Louis, 1904) H. Poincaré presented a clear formulation of the relativity principle. "According to the principle of relativity the laws of physical phenomena must be the same for a stationary observer as for an observer carried along in a uniform motion of translation; so that we have not and can not have any means of

---

[1] However, the Lorentz contraction hypothesis contradicts the principle of relativity. Lorentz and Fitzgerald believed in an absolute state of rest which pertains to all moving bodies. The shortening of moving meter sticks is here not a reciprocal effect as in the theory of relativity; and, in principle, it could serve to determine an absolute reference frame at rest. But this concept contradicts all experience.

discerning whether or not we are carried along in such a motion."[1]
However, Poincaré's argumentation is different from Einsteins. He
was familiar with Lorentz's work of 1904, and in his view the principle
of relativity should be "explained", somehow as Lorentz attempted
to give a dynamical explanation of the Lorentz contraction. On the
other hand, Poincaré also pointed out the large number of arbitrary
hypotheses in Lorentz theory. In addition he suggested that a "new
mechanics" ought to be developed in place of Newtonian mechanics.
However, he did not formulate the new mechanics; it was for him only
a program. With his discovery that the Lorentz transformations form
a group, Poincaré made a lasting contribution to the theory of relativity. In his work of 1905 he established the terms "Lorentz transformation" and "Lorentz group" and demonstrated the invariance against
Lorentz transformations of the vacuum Maxwell equations. He succeeded
in deriving the Maxwell equations from an invariant action principle.
Poincaré also managed to find an interpretation of the Lorentz transformations in terms of rotations in a four-dimensional Euclidean space
with coordinates x, y, z, ict.

But the crucial contribution came from Einstein[2] in his famous
paper "Zur Elektrodynamik bewegter Körper" (Electrodynamics of moving
bodies), Ann. Physik $\underline{17}$, 891-921 (1905). As we know today, one may
assume that Einstein was not familiar with Lorentz's 1904 work nor
with Poincaré's 1905 paper.[3] Einstein finds that it is not necessary
to assume the existence of an ether at rest. His starting point is

---

[1] English translation in H. Poincaré, The Value of Science (Dover
Publications, New York 1958).

[2] An important contribution to the scientific, philosophical, and
historical analyses of what Einstein did or did not accomplish or
assert in 1905 has been given by A. I. Miller, Albert Einstein's
Special Theory of Relativity, Emergence (1905) and Early Interpretation (1905-1911) (Addison-Wesley, Reading 1981).
For a scientific biography of Albert Einstein we refer the reader to
the excellent book by A. Pais, "Subtle is the Lord..." The Science
and the Life of Albert Einstein (Oxford University Press, Oxford
1982).

[3] G. Holton, Am. J. Phys. $\underline{28}$, 627 (1960).

that all experiments aimed at observing a motion relative to the ether failed.[1] This then justifies one to demand that the equivalence of reference frames in uniform motion (a valid statement in Newtonian mechanics) should have _general_ validity. (This statement is the "Einstein principle of relativity".) We must emphasize that in Einstein's work the principle of relativity has a fundamental axiomatic role in the theory. Consequences derived from it should be confronted with experiments.

The second assumption of Einstein is that the light velocity in vacuum has the same value in all frames of reference that are in uniform relative motion. The light velocity is a universal constant, i.e. it is independent of the velocity of the light source which is in uniform motion relative to an observer. The assumption of reference frames in uniform relative motion is important, because if one allows for accelerated systems, the light velocity looses its absolute character. Einstein does not provide a detailed justification of his two postulates, and restricts himself to ascertain that these assumptions are universally verified by facts. These two statements are only in apparent contradiction to each other because of the traditional and unfounded assumption that simultaneity of two events has an absolute meaning. In Newtonian physics one could (in a Gedankenexperiment) overtake a ray of light. Because of the simple additivity of velocities, the light velocity would be essentially different for observers in frames of reference that are in relative motion. This can be avoided if one gives up the notion of absolute time. In this manner also the apparent contradiction between the two postulates disappears; and thus, these postulates permit the construction of a non-contradictory theory of the electrodynamics of moving bodies.

Einstein analyses the definition of simultaneity in terms of measurement procedures. If one gauges the clocks applied for the measurement of time with the use of light signals, one finds that events which are simultaneous in a reference frame K, are no longer simulta-

---

[1] The Michelson-Morley experiment is not mentioned by Einstein.

neous in a system K' which is in uniform motion relative to K. One must give up the relation t'=t, which would hold in Newtonian physics. The postulates noted above are used to derive from them the Lorentz transformation; and the lengthwise contraction of measuring rods, as well as the time dilatation of clocks, is discussed. In Einstein's theory, notably, length contraction is not connected to some particular model of forces holding matter together; rather, it arises directly from the definition of length. Likewise, time dilatation arises from the definition of time. Einstein, then derives the new velocity addition law; this does not permit the overtaking of a light ray. The transformation laws for the electromagnetic field are demonstrated; the Doppler effect and the aberration of light is explained; the dynamics of an electron is described; the motion of an electron in a constant electric and magnetic field is analysed. But the famous equation $E_o = mc^2$ is not yet explicitly stated; it appears first in later researches, cf. the paper entitled "Ist die Trägheit eines Körpers von seinem Energieinhalt abhängig?" (Does the inertia of a body depend on its energy content?), Ann. Physik 18, 639–641 (1905).

In summary, one may say that the theory of special relativity was initiated by H.A. Lorentz, its physical foundations and its physical content was shown by A. Einstein, and its mathematical structure was made clearest by H. Poincaré.

Looking at it from a contemporary viewpoint, one would think that the new ideas of Einstein must have had the general effect of a revelation, as the slashing of the Gordian Knot. In actual facts, their effect on thinking came only in isolated steps and with delay. To start with, Einstein's ideas surely did not lead to a significant number of other publications in the topical area. This happened only about four years later. Max Planck (1858–1947) was apparently the first to recognize immediately the importance of Einstein's researches. Already in 1906, Planck presented the Lagrangian of relativistic mechanics in the form as used today.

Poincaré himself later came to doubt the principle of relativity. Influenced by the earlier experimental findings of W. Kaufmann on the specific charge of fast-moving electrons, Poincaré wrote this in 1906: "The principle of relativity may well not have the rigorous value which has been attributed to it".[1] Up to his death in 1912 he often discussed the principle of relativity in his writings, without, however, acknowledging the contributions of Einstein. On the other hand, H.A. Lorentz gradually came to accept relativity theory.

The next significant contribution to the theory of relativity came from Hermann Minkowski (1864-1909) who gave it a covariant formulation. He used a four-dimensional space-time continuum and showed that Einstein's theory may be expressed especially simply in the language of the pseudo-Euclidean geometry (Minkowski space). At the Meeting of the German Scientists and Medical Doctors (Cologne, 1908) Minkowski gave a popular talk on "Space and Time" which was very well received and which led to wider acceptance of relativity theory.[2] In his memories Max Born (1882-1970) says the following regarding his teacher in those days H. Minkowski: "He told me later that it came to him as a great shock when Einstein published his paper in which the equivalence of the different local times of observers moving relative to each other was pronounced; for he had reached the same conclusions independently but did not publish them because he wished first to work out the mathematical structure in all its splendour. He never made a priority claim and always gave Einstein his full share in the great discovery."[3]

The great experimentalist, Albert Michelson (1852-1931) could not come to terms with the theory of relativity. According to Einstein's recollection, at their only personal meeting in 1931, Michelson expressed toward him a certain regret that his own work started this "monster".[4]

---

[1] Quotation from S. Goldberg, Am.J.Phys. <u>35</u>, 934(1967).

[2] H. Minkowski, Phys.Z. <u>10</u>, 104(1909); also reprinted in the collection "The Principle of Relativity" (trans. W. Perrett and G.B. Jeffery), l.c.

[3] M. Born, My Life (Scribner, New York 1978), p. 131.

[4] G. Holton, Am.J.Phys. <u>37</u>, 968 (1969).

Now we come to the question whether the Michelson-Morley experiment was indeed the foundation and take-off point of Einstein's theory of relativity, as it is claimed in many textbooks. Certainly, in Einstein's 1905 work neither is the experiment mentioned nor is there a reference to other sources of literature. In a letter from 1954, Einstein, who definitely appreciated the importance of the Michelson experiment, makes the following statement: "In my own development Michelson's result has not had a considerable influence. I even do not remember if I knew of it at all when I wrote my first paper on the subject (1905). The explanation is that I was, for general reasons, firmly convinced that there does not exist absolute motion and my problem was only how this could be reconciled with our knowledge of electro-dynamics. One can therefore understand why in my personal struggle Michelson's experiment played no role or at least no decisive role."[1]

But Einstein's contemporaries were much amazed by the outcome of the Michelson experiment. It was a blow to the ether-hypothesis. However, nowadays it is an unnecessary roundabout to follow the historical path with all mistakes, in course of which one first introduces the unrealistic concept of an ether, only to eliminate it in the end. Besides, to do so, it would be necessary to take into account other experiments as well. For this reason, too, and not only from a historical point of view, one cannot take the Michelson experiment alone as the crucial experiment. One cannot deduce the theory of relativity from the Michelson experiment.

In his lecture delivered in honour of Max Planck's sixtieth birthday (1918)[2] Einstein makes a more general comment regarding this

---

[1] The full text of the letter to Davenport may be found in the quoted paper of G. Holton (1969). For more details see also the Chapter "Einstein, Michelson, and the 'Crucial' Experiment" in the historical studies by G. Holton, Thematic Origins of Scientific Thought; Kepler to Einstein (Harvard University Press, Cambridge, Mass. 1973).

[2] This lecture is reprinted with the title "Principles of Research" in A. Einstein, Essays in Science (Philosophical Library, New York, 1934).

problem. He says: "The supreme task of the physicist is, then, to search for those most general elementary laws from which one can gain the world-picture ("Weltbild") by pure deduction. No logical path leads to these elementary laws; but only intuition, supported by insight into experience, can lead to them."

Accordingly, we shall employ only Einstein's principle of relativity to derive therefrom the theory of special relativity.

Chapter 2

PHYSICAL AND CONCEPTUAL
FOUNDATIONS OF THE THEORY OF
SPECIAL RELATIVITY

2.1 The Hypotheses of Newtonian Mechanics

The centerpiece of Newtonian mechanics is the Newtonian equation of motion,

$$\frac{d}{dt}\left(m\,\frac{d\vec{r}}{dt}\right) = \vec{F} \qquad (2.1)$$

but, besides the well known Newtonian axioms, also a number of additional assumptions enter into the formulation. Often these are not explicitly mentioned, since, apparently, they are taken selfevident.[1] However, the theory of special relativity leads to a change of some of these hypotheses. Therefore, we briefly give an account of these tacit hypotheses.

In three-dimensional Euclidean space one introduces orthogonal Cartesian coordinates which are used to describe the position of a particle that moves in accord with equation (2.1). Space is isotropic and homogeneous. The relevant vectors obey the rules of vector algebra

---

[1] Regarding the "common sense", which one might invoke here, one should quote Einstein's poignant statement: "Common sense is that layer of prejudices laid down in the mind prior to the age of eighteen."

and vector analysis. Among all possible coordinate systems in relative motion to each other, these is a distinguished class. In this class of coordinate systems, called inertial systems the Galilean law of inertia holds valid. (This is Newton's Lex Prima.) Only in an inertial system (also characterizable as an unaccelerated system) will a body free of applied forces ($\vec{F}=0$) stay at rest or preserve its motion with constant velocity. It is advantageous to formulate physical laws (such as Newton's equation of motion) relative to an inertial system, because in this case the laws take their simplest forms. Thus, for example, the simple form (2.1) of Newton's equation of motion is valid only in an inertial system. A good example of an inertial system is a spaceship with shut-off engines and without any self-rotation. Any other system that has, relative to the spaceship, an acceleration, is not an inertial system. In such accelerated systems a relation, rather similar to equation (2.1) holds true. Alongside the applied force $\vec{F}$ virtual forces will appear too (e.g. centrifugal force, Coriolis force). A coordinate system rigidly connected with the Earth is not really an inertial system, because the Earth rotates and moves around the Sun. Even the Sun moves along an orbit around the center of the Galaxy. However, in praxis one may neglect the small effects of the Earth's rotation and its orbital motion, or the motion of the Sun ("fixed stars"). Thus, a coordinate system which is attached to the Earth or the Sun (or fixed stars) may be considered an (approximate) inertial systems. Examples of situations where one must reckon with the additional inertial forces are the Foucault pendulum or the deviation of a body in free fall from a great hight due to the Earth's rotation.

The second assumption concerns time. In Newtonian mechanics one introduces an absolute time scale which may be used in all coordinate systems that are in mutual relative motion. Consequently, the statement of simultaneity of two events also has an absolute meaning. The concept of simultaneity is important, because, in order to measure the length of a meter stick, one must mark off simultaneously the beginning and the end point of the stick in a system at rest. This is essential for the definition of the length of a moving body. In

Einstein's relativity theory the hypothesis of absolute time is given up. In addition, in Newtonian mechanics, it is assumed that, when passing to an other coordinate system, the mass occurring in the equation of motion (2.1) is unchanged, i.e. that mass is an invariant quantity. But in the theory of relativity mass becomes a velocity dependent entity, and the invariant relativistic mass is defined in a different manner.

2.2   The Galilean Relativity Principle and its Limits

We can now easily understand that, in Newtonian mechanics, all inertial systems in mutual uniform motion are equivalent. It is not possible to identify an inertial system at absolute rest with the aid of purely mechanical processes. To prove this statement, consider two coordinate systems K and K' which move with the relative velocity $\vec{v}$ and which coincide at t=0. The coordinates of a point P, relative to K' and K, respectively, are connected to each other via a Galilean transformation

$$\vec{x}' = \vec{x} - \vec{v}t$$
$$t' = t \tag{2.2}$$

Since the velocity $\vec{v}$ does not depend on time, and since (according to the hypothesis of absolute time) t=t' holds, the accelerations in the two coordinate systems are equal,

$$\frac{d^2\vec{x}'}{dt'^2} = \frac{d^2\vec{x}}{dt^2} \quad .$$

Since, furthermore, the mass is a constant independent of the coordinate system, it follows that

$$\vec{F} = m\frac{d^2\vec{x}'}{dt'^2} \quad .$$

But the right-hand side of this equation may be taken for the definition of the force in the system K', i.e.

$$\vec{F}' = m \ \frac{d^2 \vec{x}'}{dt'^2} \ .$$

Thus, one concludes that the laws of mechanics are the same in all reference frames which arise from an inertial system through Galilean transformations. This is the relativity principle of Newtonian mechanics. Accordingly, so long as one uses purely mechanical experiments, it is not possible to distinguish a particular inertial system as one "at rest".

The Galilean transformations form a group. Here we only exhibit the group property of the group elements arising from the composition of two group elements specified by the velocities $\vec{v}_1$ and $\vec{v}_2$. If one performs in succession the transformations

$$\vec{x}' = \vec{x} - \vec{v}_1 t \ , \quad t' = t$$
$$\vec{x}'' = \vec{x}' - \vec{v}_2 t \ , \quad t'' = t'$$

one again obtains an element of the group, viz.

$$\vec{x}'' = \vec{x} - \vec{v} t \ , \quad t'' = t \ ,$$

which is specified by the velocity $\vec{v} = \vec{v}_1 + \vec{v}_2$. This is the velocity addition law valid in Newtonian mechanics.

Quantities which are not altered by any of the transformation of a group are called invariants of the transformation group. Thus, for example, the mass introduced above is an invariant of the Galilei group. As we saw earlier, applying Galilean transformations do not change the form of Newton's law and we say that it is covariant (or form-invariant) under these transformations. Thus, the Galiean relativity principle (or, rather, invariance principle) may be formally expressed by demanding that the laws of mechanics (or possibly those of some other theory) should be covariant under the transformations of the Galilei group.

Actually, this invariance principle is still restricted to Newtonian mechanics: the laws of electrodynamics etc. are not affected. But physics is concerned about general principles. Could the Galilean invariance principle represent an universal law? The answer is no.

In order to convince ourselves of this, let us examine the following two, directly related questions: 1. Is Newtonian mechanics valid at high velocities? 2. Is electrodynamics covariant under Galilean transformations?

1) So as to test mechanics at high speeds, we study an experiment in which electrons are given very high velocities via the application of a high potential (MeV) in a linear accelerator. After the electrons (with charge e) traversed in the accelerator a potential difference φ, they have the kinetic energy $E_{kin} = e\varphi$[MeV]. A subsequent drift section (length l) is traversed by the electrons during the time interval t which is measured as follows. Two electrodes, placed at the beginning and at the end of the drift section, receive electrical pulses when the charged particles enter and exit, respectively. These two pulses are conducted (through cabels of equal length) to a cathode-ray oscilloscope. The transit time of the signals from the points of measurement is equal and therefore one can determine the passage time t of the electrons from the pulse distance seen on the oscilloscope. From the relation v=l/t we then obtain the electron velocity. This experiment was performed[1] at the energies E_{kin}=0.5, 1.0, 1.5, 4.5 and 15 MeV, as illustrated in Fig.2.1. In addition, with a calorimetric measurement of the electron energy at the end of the drift chamber one verified that the accelerated electrons indeed had the calculated kinetic energy $e\varphi$.

The dotted straight line in Fig.2.1 represents the linear relation $v^2 = 2E_{kin}/m_e$ which one would expect from Newtonian mechanics. The slope of this straight line is $2/m_e$. Thus, for particles which have higher mass than electrons, it would be flatter. Now, the measurements (solid line in Fig.2.1) indicate a drastic deviation from the Newtonian prediction at higher energies (i.e., high velocities). Furthermore one finds that the electrons cannot achieve arbitrary high velocities, and that, clearly, the limiting speed equals the velocity of light in vacuum ($c=3\times10^8$m/sec). This is also verified by the measurement at 15 MeV (not shown in the figure), and even more impressively

---

[1] Additional details of this experiment may be found in W. Bertozzi, Am.J.Phys. 32, 551(1964).

16

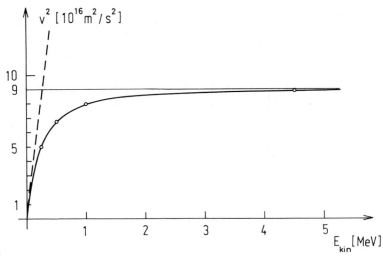

Fig. 2.1

demonstrated by new measurements at energies three orders of magnitude higher, in the range of $15-20 \times 10^3$ MeV.[1] In this experiment the time of flight of high-energy electrons for a flight path of about 1 km was compared with that of light. By using a time-of-flight technique with 1 psec sensitivity no significant difference in the velocities of light and electrons was observed to within ~ 2 parts in 10^7. One should note that, using the units on the abscissa as in Fig. 2.1, the result of this experiment ought to be entered at about 280 m from the origin.

Thus, even extremely high energies cannot be used to give super-luminal velocities to particles with as little mass as the electrons have. In contrast, according to Newtonian mechanics, electrons with kinetic energy 20×10^3 MeV should be 283 times faster than light.

In summary, we found that Newtonian mechanics loses its va-lidity at high velocities, and that the speed of light in vacuum, $c = 3 \times 10^8$ m/sec, is a limiting velocity which cannot be surpassed.

---

[1] These measurements were performed at the Stanford Linear Accelerator Center (see Z.G.T. Guiragossián et al., Phys. Rev. Lett. 34, 335 (1975).

All experiments indicate that c is the maximal speed with which effects can propagate. It determines the minimal time interval necessary for the passage of a physical change from its point of origin to the point of its effect. Since the propagation velocity of interactions cannot exceed c, there are no instantaneous effects (action-at-a-distance) in nature. This could happen if infinite velocities existed. However, Newtonian mechanics of mass points are based on the assumption of action-at-a-distance. Therefore, strictly speaking, Newtonian mechanics is not exact, but it is a very good approximation as long as the relevant velocities v are small compared to the light velocity c, i.e., if $v^2/c^2 \ll 1$ holds. Now we turn to the second question.

2) Is electrodynamics invariant under Galilean transformations? One should not expect this to hold, because, according to the simple velocity addition law, one could achieve a light velocity c+v or c-v, resp., in moving reference systems. Then c ($c=1/\sqrt{\varepsilon_o \mu_o}$ in the MKSA-system) would not be the maximal propagation velocity, and electromagnetic effects in inertial systems that are in relative motion, would not be the same.

One can also verify by calculation that the wave equation for electrodynamics, such as valid for the scalar potential $\varphi(x,t)$,

$$\left(\vec{\nabla}^2 - \frac{1}{c^2} \frac{\partial^2}{\partial t^2} \right) \varphi = 0$$

is not covariant under Galilean transformations. It is the Faraday induction term which destroys the invariance. If one omits in the Maxwell equations the induction term, one obtains for the electric and magnetic fields $\vec{E}$ and $\vec{B}$ the differential equations

$$\text{div } \vec{E} = \varrho \text{ , curl } \vec{E} = 0$$
$$\text{div } \vec{B} = 0 \text{ , curl } \vec{B} = \vec{j} + \frac{\partial \vec{E}}{\partial t}$$

which are covariant under Galilean transformations. Here $\varphi(\vec{x},t)$ and $\vec{j}(\vec{x},t)$ denote the charge- and current-density, resp. The units used in these equations differ from the rationalized Gaussian system only in as much that the magnetic field strength $\vec{B}$ has c times its value measured in the Gaussian system. Then the omitted induction term is $(1/c^2)\, \partial\vec{B}/\partial t$, and c does not occur in Ampère's law. Since the additional Maxwell term $\partial\vec{E}/\partial t$ is present, electric charge obeys the equation of continuity. While these equations are covariant under Galilean transformations, they do not agree with experience. Since they do not contain the induction term, the phenomena related to induction, so essential for a usable theory of electromagnetic fields, are not encompassed.[1]

The fact that Newtonian mechanics is invariant under Galilean transformations while the Maxwell equations are not, leads one naturally to the following possible viewpoints:

1. There holds a principle of relativity in mechanics, but none in electrodynamics. In this case one could discerne with electromagnetic experiments a privileged reference frame in which the ether is at rest. In such manner, we would be led back to the concept of a system at absolute rest.

2. Some principle of relativity holds for both the theories, but the Maxwellian laws of electrodynamics are not correct. In this case it should be possible to conduct experiments which demonstrate the deviations from the Maxwellian theory and one would be compelled to reformulate the laws of electrodynamics.

---

[1] The law of induction was discovered by Faraday in 1831. This also happens to be Maxwell's year of birth. It is quite possible to think that if Maxwell had lived before 1831, he perhaps would have formulated a Galilean invariant electrodynamics. In that case, the subsequent discovery of Faraday would have given the physicists the following choice: either abandone the relativity principle for electrodynamics (ether hypothesis), or generalize it (the theory of special relativity). Thus, if one first introduces the displacement current and the Galilean invariance principle, and discusses the induction term destroying Galilean invariance only later, one may build an interesting approach to electrodynamics. Further discussion of an electrodynamics covariant under Galilean transformations may be found in the literature. Cf. M. Le Bellac and J.M. Levy-Leblond, Nuovo Cimento 14B, 217 (1973); M. Jammer and J. Stachel, Am. J. Phys. 48, 5 (1980).

3. Some principle of relativity holds both for mechanics and electrodynamics, but the Newtonian laws of mechanics are not correct. Then one should be able to find experimental discrepancies
in Newtonian mechanics, and mechanics ought to be reformulated
correspondingly.

We already noted that experiments with high-velocity electrons
indeed reveal deviations from Newtonian mechanics. We also indicated
that, while one can formulate a theory of the electromagnetic field
which obeys Galilean relativity, but this theory is not in accord
with facts. On the other hand, we know that the laws of Maxwellian
electrodynamics are correct and do not require any alteration. Thus,
only the last of the three possibilities can hold. The principle of
relativity valid for both mechanics and electrodynamics cannot be
the Galilean principle. We must replace the Galilean transformations
by some other kind of transformations; and one must reformulate mechanics in such a way that it obeys the new relativity principle and
that at low velocities ($v^2 \ll c^2$) it approximates Newtonian mechanics.

2.3   The Einsteinian Principle of Relativity

In summary, experience shows that there must hold a principle
of relativity which is more general than the Galilean principle. Accordingly, it is impossible to determine an inertial reference frame
at absolute rest by using any (not only mechanical) physical phenomena. Inertial systems are fully equivalent, and the laws of nature
can be written down in a form which is covariant under the (yet to
be determined) transformations which connect different inertial systems. Consequently, the laws have the same form in all inertial systems. This is Einstein's principle of relativity.

In particular, it follows that, similarly to laws of nature,
the maximal propagation velocity of interactions must be the same
in all inertial sytems. Thus, the limiting speed c is a universal
constant. (This is Einstein's second postulate.) Otherwise one could
get in a suitably chosen inertial system a larger (or smaller) propa-

20

gation velocity. This circumstance could be used to distinguish such a particular inertial system, contradicting the equivalence of all inertial systems as stated by the universal principle of relativity.

But now we must abandon the notion of absolute time which leads to the Galilean law of velocity addition. Since, strictly speaking, there are no instantaneous effects, one must take into account the finite propagation velocity of an effect (such as a light signal) when one synchronizes clocks at different locations.

Thus, the principle of relativity leads to the result that time cannot be absolute. In different reference systems time runs differently. Passing from one to another reference system, time (like position) must be transformed: $\{x,y,z,t\} \rightarrow \{x',y',z',t'\}$. Therefore, specification of a time interval makes sense only if one also specifies the reference frame to which the statement refers. For example, if two events are simultaneous in one inertial sytem, they will not be simultaneous in some other systems. We illuminate this point with two examples.

According to the Doppler effect, the light from a source that moves toward an observer assumed at rest will suffer a blue shift, i.e. the wavelength appears reduced. So as to keep the universal light velocity c constant, the frequency must increase in accord with the relation $c= \lambda \nu$. But frequency is a measure of time (dimension t^{-1}), i.e., time runs differently in the moving system than in the one at rest. Correspondingly, the notion of simultaneity seizes to be an absolute concept and is meaningful only with reference to a specified inertial system.

As a second example, let us consider an inertial system K' which, relative to K, moves toward the right (see Fig.2.2). Let C be a point on the x' axis half-way between A and B. Seen from system K', a light signal emitted from C into both directions will reach the points A and B simultaneously. However, this statement

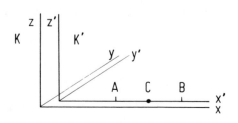

Fig. 2.2

does not hold true for an observer at rest in K. In the system K the signal's velocity is also c. The point A moves toward the signal (relative to the system K), and the point B moves away from the signal which was sent to it. Consequently, in system K the signal reaches the point A sooner than B.

The concept of simultaneity enters into the measurement of length. For example, in order to measure the length of a rod, one must mark simultaneously the positions of its endpoints. Therefore, one should expect that, according to the universal principle of relativity, the length of a rod also loses its absolute meaning. We shall come back to the problem of length contraction later. But first we will find the transformations which correspond to the Einsteinian principle of relativity.

2.4   The Lorentz Transformation

In the present derivation of the Lorentz transformations we shall assume only geometrical properties of space and the validity of the Einsteinian principle of relativity. The Lorentz transformations so deduced contain a universal constant, which is finite and has the character of a limiting speed. This holds independently of its numerical value which does not follow from the theory but has to be determined experimentally. Experience shows that this invariant limiting speed must be identified with the velocity of light in vacuum.

On the other hand, if, to begin with, one adds the constancy of light velocity as an explicit postulate, the Lorentz transformations can be obtained more speedily.[1]

---

[1] The literature on the derivation of the Lorentz transformations is rather voluminous. Suffice here to refer only to the following more recent works: G. Süßmann, Z. Naturf. A24, 495(1969); J.-M. Levy-Leblond, Am. J. Phys. 44, 271 (1976).

2.4.1 Deduction of the Lorentz Transformation from the Principle of Relativity

We start with the following assumptions on the symmetries of space and of physical phenomena:

1. Space is isotropic, i.e., all spatial directions are equivalent.

2. Space and time are homogeneous, i.e., no point of space or time is distinguished. The origin of the coordinate system may be chosen arbitrarily, without having an effect on the measuring devices for space and time.

3. The universal principle of relativity holds, i.e., all inertial systems are equivalent.

These assumptions determine the transformations between the coordinates of the inertial systems K and K' up to a constant. In fact, the first two assumptions form a part of the complete definition of an inertial system.

Because of the homogeneity of space and time, the transformations must surely be linear. Otherwise the choice of the origin of the coordinate system would not be arbitrary and certain points or regions of space would be distinguished. Under linear transformations the straight lines of inertial motion in K transform into straight lines in K', as one expects. In contrast, nonlinear transformations would lead to accelerations in K', which were not present in K.

a) Let the origins of the systems K and K' coincide at the initial time t=0. Let system K' move in the direction of the positive x axis with the constant velocity v. Then, after a time interval t in K the origin of K' (given by x'=y'=z'=0) takes on the position coordinates x=vt, y=z=0. Then the linear transformations from K to K' may be written as follows:

$$
\begin{aligned}
x' &= \gamma(v)\,(x - vt) \\
y' &= \alpha(v)\,y \\
z' &= \alpha(v)\,z \\
t' &= \mu(v)\,t + \epsilon(v)\,x
\end{aligned}
\qquad (2.3a)
$$

Here the constants $\gamma, \alpha, \mu, \varepsilon$ are independent of the coordinates, but they may depend on the relative velocity v. Because of the isotropy of space, the coefficients in the second and third equation ($\alpha(v)$) must be equal. In the equation for t', y and z do not occur. Otherwise clocks arranged symmetrically to the x axis (say at +y,-y or +z,-z) would give different readings in the system K', contradicting the isotropy of space. Now we proceed to determine the four unknown coefficients $\gamma, \alpha, \mu, \varepsilon$.

b) Since isotropy of space means that no direction is distinguished, the transformation formulae (2.3a) must not change if the direction of the x axis, x' axis and the relative velocity v are simultaneously reversed. That is,

$$-x' = \gamma(-v)(-x+vt)$$
$$-y' = -\alpha(-v)y$$
$$z' = \alpha(-v)z$$
$$t' = \mu(-v)t - \varepsilon(-v)x \quad .$$

In order to assure that K and K' remain right-handed systems, we also changed the signs of y and y'. Comparison with (2.3a) shows that

$$\gamma(-v) = \gamma(v), \quad \alpha(-v) = \alpha(v), \quad \mu(-v) = \mu(v)$$

but

$$\varepsilon(-v) = -\varepsilon(v) \quad .$$

The odd function $\varepsilon(v)$ may be conveniently replaced by function $\eta(v)$ which is even in v:

$$\varepsilon(v) = -\frac{v}{\eta(v)}\mu(v)$$

Now the transformation equations (2.3a) read:

$$x' = \gamma(v)(x-vt)$$
$$y' = \alpha(v)y$$
$$z' = \alpha(v)z \quad \quad (2.3b)$$
$$t' = \mu(v)\left(t - \frac{v}{\eta(v)}x\right)$$

c) According to the relativity principle the inertial systems K and K' are completely equivalent. Therefore the transformation equations for the transition from K' to K must have the same form as those for the transition from K to K'. But the inverse transformation (K' → K) must contain v with an opposite sign, because, as seen from K', the system K moves in the direction opposite to the original direction in which K' moves relative to the system K that was considered "stationary" to start with. Thus, for the transition from K' to K one must replace the unprimed coordinates with the primed coordinates and one must reverse the sign of v:[1]

$$x = \gamma(v)(x' + vt')$$
$$y = \alpha(v)y'$$
$$z = \alpha(v)z' \qquad\qquad (2.3c)$$
$$t = \mu(v)\left(t' + \frac{v}{\eta(v)}x'\right)$$

Here we already took into consideration that the coefficients $\gamma, \alpha, \mu, \eta$ must be even functions of v. If one substitutes $y' = \alpha(v)y$ into the second equation, one immediately gets $\alpha^2(v) = 1$, i.e., $\alpha = \pm 1$. In order to assure that, for $v \to 0$ the transformation goes continuously to the identity transformation, one must take the positive sign, i.e, we have

$$\alpha = 1$$

In order to determine the remaining coefficients, let us multiply the first equation in (2.3b) with μ, the last one with $v\gamma$, and add them:

$$\mu x' + v\gamma t' = \mu\gamma\left(1 - \frac{v^2}{\eta}\right)x$$

Solving for x gives

$$x = \frac{x'}{\gamma\left(1 - \frac{v^2}{\eta}\right)} + \frac{vt'}{\mu\left(1 - \frac{v^2}{\eta}\right)}$$

---

[1] This "relativistic replacement" will also be useful later on.

and, comparing with

$$x = \gamma (x' + v t')$$

as given in (2.3c), it follows that

$$\gamma = \frac{1}{\gamma (1 - \frac{v^2}{\eta})} \quad , \quad \gamma = \frac{1}{\mu (1 - \frac{v^2}{\eta})} \quad ,$$

hence

$$\gamma = \mu \quad , \quad \gamma^2 = \frac{1}{1 - \frac{v^2}{\eta(v)}} \quad .$$

Because of the required continuity of the passage into the unit element when $v \to 0$, only the positive root makes sense, that is

$$\gamma = \frac{1}{\sqrt{1 - \frac{v^2}{\eta(v)}}} \tag{2.4}$$

Thus, the parameters of the transformation are determined, with the exception of one. The still unknown quantity η must be either a constant or an even function of v, and it has the dimension of a squared velocity.

d) So as to obtain more information on η, we again use the principle of relativity, according to which the above transformations must be universal, i.e., they should hold for transitions between arbitrary inertial systems. Thus, if one transforms from K to K' and subsequently to K'', the formulae connecting K and K'' must have the same form as those, given above. Performing two Lorentz transformations in succession, one again obtains a Lorentz transformation. Since, in addition, the associative law is obeyed, the inverse element and the unit element exist, the Lorentz transformations form a group. Let v_1 be the velocity of K' relative to K and v_2 the velocity of K'', relative to K', in the directions of the positive x and x' axis, respectively. Then we have

$$x' = \gamma(v_1)(x - v_1 t) \quad , \quad x'' = \gamma(v_2)(x' - v_2 t')$$

$$t' = \gamma(v_1)\left(t - \frac{v_1}{\eta(v_1)}x\right), \quad t'' = \gamma(v_2)\left(t' - \frac{v_2}{\eta(v_2)}x'\right),$$

and expressing x'', t'' in terms of x, t it follows that

$$x'' = \gamma(v_2)\gamma(v_1)\left[x - v_1 t - v_2\left(t - \frac{v_1}{\eta(v_1)}x\right)\right]$$

$$t'' = \gamma(v_2)\gamma(v_1)\left[t - \frac{v_1}{\eta(v_1)}x - \frac{v_2}{\eta(v_2)}(x - v_1 t)\right]$$

(2.3d)

Now, this transformation (from K to K'') is an element of the Lorentz group so that it can be written in the form

$$x'' = \gamma(w)(x - wt)$$

$$t'' = \gamma(w)\left(t - \frac{w}{\eta(w)}x\right)$$

(2.3e)

where the velocity w stands for the group parameter defining the transformation from K to K''.

We note that for a Lorentz transformation the coefficient of x in the transformation equation for x is the same as the coefficient of t in its transformation equation (i.e., $\gamma = \mu$). Comparing the corresponding coefficients in (2.3d) it follows that

$$\gamma(v_2)\gamma(v_1)\left[1 + \frac{v_1 v_2}{\eta(v_1)}\right] = \gamma(v_2)\gamma(v_1)\left[1 + \frac{v_1 v_2}{\eta(v_2)}\right]$$

In order to satisfy this relation, it is necessary to have

$$\eta(v_1) = \eta(v_2) = const.$$

Thus, the quantity η is a universal constant with the dimension of a squared velocity. This constant must have a finite value. This is so because in the limit of $\eta \to \infty$ the above transformation formulae become the familiar Galilean transformations which do not correspond to Einstein's relativity principle and which lead to correct prediction only in the limiting case of small velocities, $v^2/c^2 \ll 1$.

    e) Even without knowing yet the value of the universal constant η, one can now discern the relativistic velocity addition law from the group property of the transformations. One starts with the composite transformation (2.3d),

$$x'' = \gamma(v_2)\,\gamma(v_1)\left(1 + \frac{v_1 v_2}{\eta}\right)\left[x - \frac{v_1 + v_2}{1 + \frac{v_1 v_2}{\eta}}\,t\right]$$

and finds that the square of the first three factors may be written as

$$\frac{\left(1 + \frac{v_1 v_2}{\eta}\right)^2}{\left(1 - \frac{v_2^2}{\eta}\right)\left(1 - \frac{v_1^2}{\eta}\right)} = \frac{1}{1 - \frac{1}{\eta}\left(\frac{v_1 + v_2}{1 + \frac{v_1 v_2}{\eta}}\right)^2}$$

Comparison of

$$x'' = \frac{1}{\sqrt{1 - \frac{1}{\eta}\left(\frac{v_1 + v_2}{1 + \frac{v_1 v_2}{\eta}}\right)^2}}\left[x - \frac{v_1 + v_2}{1 + \frac{v_1 v_2}{\eta}}\,t\right]$$

with the value of the velocity w corresponding to the Lorentz transformation from the system K to K'' (given by equ.(2.3e)) yields

$$w = \frac{v_1 + v_2}{1 + \frac{v_1 v_2}{\eta}} \tag{2.5}$$

This is Einstein's addition theorem for parallel velocities v_1 and v_2. It becomes the simple addition law of Newtonian mechanics if one can neglect the second term in the denominator in comparison to 1. This is to be expected since for small velocities $v_1 v_2 / \eta \ll 1$ Newtonian mechanics is valid to a high accuracy. This indicates that the numerical value of the universal velocity $\sqrt{\eta}$ must be large compared to the velocities occuring in classical mechanics.

But the most important consequence of the addition law (2.5) is that, with $\eta > 0$, the universal velocity $\sigma = \sqrt{\eta}$ cannot be exceeded. In other words, the parameter $\sigma = \sqrt{\eta}$, whose numerical value is not yet determined, represents a limiting velocity.

In order to show this , we assume that a body in the system K' moves with the velocity $v' = \sigma - \alpha'$, $\alpha' > 0$, in the direction x'. Let the system K' move relative to K with the velocity $v = \sigma - \alpha$, $\alpha > 0$, in the direction of x. Then, according to the addition law (2.5), the velocity of the body relative to K is

$$w = \frac{2\sigma - \alpha' - \alpha}{2\sigma - \alpha' - \alpha + \frac{\alpha\alpha'}{\sigma}}, \quad \sigma < \sigma$$

From here we see that the velocity cannot exceed σ. Even if a process propagates in system K' with the velocity σ, the addition theorem (3.5) tells us that we have the same velocity for this limiting speed in the system K:

$$w = \frac{v + \sigma}{1 + \frac{v}{\sigma}} = \sigma .$$

Once again this expresses the fact that σ is a universal velocity, i.e. it is the same in every inertial system.

The numerical value of σ must be determined experimentally. According to all measurements up to date, especially the experiments regarding the limiting velocity as was discussed in Sec.2.2, σ cannot be distinguished from the velocity of light in vacuum. Therefore one assumes that the relativistic limiting speed σ and the velocity of light c are identical. Then the Lorentz transformations, which follow from the Einsteinian relativity principle, assume the form

$$
\begin{aligned}
x' &= \gamma (x - v t) \\
y' &= y \\
z' &= z \\
t' &= \gamma (t - \frac{v}{c^2} x) , \quad \gamma = \frac{1}{\sqrt{1 - \frac{v^2}{c^2}}}
\end{aligned}
\qquad (2.6)
$$

and the addition theorem reads

$$W = \frac{V_1 + V_2}{1 + \dfrac{V_1 V_2}{c^2}} \tag{2.7}$$

Regarding the obvious question as to why the maximal propagation speed σ is precisely the light velocity, we make the following comment. As we shall see later on, it follows from the Lorentz transformation that only those particles can move with the limiting speed σ occuring in the transformation equation which have a vanishing invariant mass. According to the nowadays generally accepted theory of the electromagnetic field, one is convinced that this holds for the particle corresponding to light, i.e. for the photon. But, in principle, experiments do not exclude the possibility that the photon has a mass slightly different from zero. In this case, the velocity of light would be slightly less than the limiting speed σ and would depend on frequency. According to newest estimates, the experimental upper limit for the rest mass of the photon is $m_{ph} \leq 5 \times 10^{-60}$ g $= 3 \times 10^{-27}$ eV.[1]

However, presently achievable accuracies are not sufficient to measure the deviation of the light velocity from σ which would correspond to this minute photon mass.[2]

It should be mentioned here that in 1983, a new definition of the length unit meter has been adopted by the Conférence Générale des Poids et Mesures. According to this, the meter is defined as the distance traversed by light in vacuum during the time interval of 1/299 792 458 seconds. Thus the speed of light is defined to be c=299 792 458 m/sec. In consequence of the definition of the vacuum

[1] See Particle Data Group, Phys. Lett. 204 B (1988), p.13, and ibid. p.134 for further references. Earlier estimates have been discussed in detail by A.S. Goldhaber and M.M. Nieto, Rev. Mod. Phys. 43, 277 (1971).

[2] Currently the speed of light in vacuum is measured with an error limit of about 1m/sec. and has the value c = (299 792 458 ± 1.2) m/sec. (See Particle Data Group, Phys. Lett. 170B (1986), p.36). The accuracy of about 1m/sec. was achieved among others, by K.M. Evenson et al., Phys. Rev. Lett. 29, 1346 (1972).

light velocity so given, the limit of length is attached to the time unit second. The new definition of the meter is based on the progress of measuring techniques. Since today it is possible to reproduce the time standard very accurately by measuring frequencies, one can then also determine a length (such as λ) with a corresponding accuracy, if one uses the relation $c = \lambda \nu$ and the definition of c.[1]

In the above considerations leading to the limiting speed σ, we assumed that $\eta > 0$. To complete the discussion, we now explain why the mathematically possible case $\eta < 0$ must be excluded. First we note that for $v = c$ the equations (2.6) become meaningless, because in this case the denominator of the coefficient γ vanishes. Clearly, this expresses the fact that the limiting speed c cannot be <u>exceeded</u>. On the other hand, if one took $\eta < 0$, the coefficient $\tilde{\gamma} = 1/\sqrt{1 + v^2/|\eta|}$ would not lead to a limitation of velocities, so that, in principle, one could have $v_1 v_2 / |\eta| > 1$. However, in this case the velocity addition law would read

$$\tilde{w} = \frac{v_1 + v_2}{1 - \dfrac{v_1 v_2}{|\eta|}}$$

so that with $v_1 v_2 / |\eta| > 1$ one would get the physically nonsensical result that two velocities in the same direction add up to one pointing in the opposite direction.

Also the following argument may be useful. A homogeneous linear transformation $(x,t) \to (x',t')$ has two fixed straight lines, which may be either real or imaginary: $x = \pm ct$ (Lorentz transformation), $x = \pm ict$ (rotations). In the latter case, with $\eta = (ic)^2 < 0$, the transformations would be isomorphic to rotations in a two-dimensional space and hence they would be compact. But this contradicts experience, since the causal propagation of light $(x^2 = c^2 t^2)$ is compatible only with $\eta > 0$, i.e., with the noncompact Lorentz group.

---

[1] Further details may be found in B.W. Petley, Nature <u>303</u>, 373 (1983), and also P. Giacomo, Am. J. Phys. <u>52</u>, 607 (1984).

The choice $\eta=-c^2 < 0$ would also contradict the increase with velocity of the "relativistic mass" m_r of an electron, which was verified experimentally with high accuracy. As we shall see later on, from the Lorentz transformation (i.e., using $\eta=c^2 > 0$) it follows that

$$m_r = \frac{m}{\sqrt{1 - \frac{v^2}{c^2}}}$$

where m stands for the invariant mass (or rest mass) of the body.[1] If we took $\eta < 0$, then under the root sign the two terms would add (rather than substract), so that m would decrease with increasing v. But this contradicts experiment.

The established mass increase also explains the experiment discussed in 2.2, according to which electrons ($m \neq 0$) cannot be brought to superluminal velocities. Clearly, this may be interpreted by means of the increase with velocity of the inertial mass. In contrast, for $\eta=-c^2 < 0$, the inertia of moving electrons would decrease when the velocity gets higher, and one would have a conflict with experiment.

We saw, the Lorentz transformations follow directly from the relativity principle together with the symmetry properties of space and time, i.e. the homogeneity of space and time and the isotropy of space. They contain the finite universal limiting speed σ, which, in agreement with experiment may be identified with the velocity of light in vacuum. In this manner, the universality of c follows from the Einsteinian relativity principle and does not have to be postulated. In contrast, for $c \neq \sigma$ the photon would have a rest mass different from zero and c would not be universal. Thus, Einstein's postulate of the constancy of the velocity of light corresponds to the requirement that the invariant mass of the photon is zero.

If one assumes the constancy of light velocity from the outset, then the derivation of the Lorentz transformation becomes shorter, and the relativity principle itself retreats to the background.

---

[1] The experiments of V. Meyer, W. Reichart, H.H. Staub, H. Winkler, F. Zamboni, and W. Zych, Helv. Phys. Acta 36, 981 (1963) verify this mass increase with an accuracy of about 0.05%.

2.4.2 Deduction of the Lorentz Transformation from the Constancy of Velocity of Light

As before, let us assume, that K and K' are inertial systems in relative motion. Since, for small velocities, the Lorentz transformation should become the Galilean transformation x'=x-vt, one feels likely to put forward the following Ansatz:

$$x' = \gamma(v)(x - vt) ,$$

where the corrective factor γ approaches 1 for $v^2 \ll c^2$. Just as in step b) in the preceding derivation, one again finds that γ is an even function of v, $\gamma(v)=\gamma(-v)$. Neither of the systems K or K' is a privileged one, so that for the reversed transition the corresponding Ansatz holds:

$$x = \gamma(v)(x' + vt') .$$

Because of the constancy of light velocity (Einstein's second postulate) c is the same in the two systems, c'=c. Therefore, if at t=t'=0 a light signal is emitted from the common origin, then, in the system K resp. K' and after the time span t resp. t' the signal arrives to the location x=ct and x'=ct', respectively. Substituting t resp. t' into the above equations gives

$$x' = \gamma x \left(1 - \frac{v}{c}\right)$$

$$x = \gamma x' \left(1 + \frac{v}{c}\right) .$$

Multiplying these equations with one another, the coordinates can be eliminated and one finds that

$$\gamma(v) = \frac{1}{\sqrt{1 - \frac{v^2}{c^2}}}$$

and therefore,

$$x' = \frac{x - vt}{\sqrt{1 - \beta^2}} \quad , \quad \beta = \frac{v}{c} \; .$$

Using then $t'=x'/c$ and $t=x/c$, one can obtain from these relations the transformation formula for time:

$$t' = \frac{\frac{x}{c} - \frac{v}{c^2} ct}{\sqrt{1 - \beta^2}} = \frac{t - \frac{v}{c^2} x}{\sqrt{1 - \beta^2}} \; .$$

The above are precisely the Lorentz transformations (2.6). One only has yet to determine the coefficient α for the transformation of the y and z coordinates. This can be done as before, using the principle of relativity, and one finds $\alpha = 1$.

2.4.3 The Lorentz Transformation for Arbitrary Relative Velocities

In order to find the Lorentz transformation for the case where $\vec{v}$ has an arbitrary direction, one must decompose the position vector $\vec{x}$ into a vector parallel to $\vec{v}$, $(\vec{x}_{\parallel} \parallel \vec{v})$, and into a vector orthogonal to $\vec{v}(\vec{x}_{\perp} \perp \vec{v})$. For each component of the vector $\vec{x}_{\parallel}$ the already known transformation formula holds, while the components of $\vec{x}_{\perp}$ remain unchanged. Thus,

$$\vec{x}_{\parallel}' = \gamma (\vec{x}_{\parallel} - \vec{v}t)$$

$$\vec{x}_{\perp}' = \vec{x}_{\perp}$$

$$t' = \gamma \left(t - \frac{\vec{v} \cdot \vec{x}_{\parallel}}{c^2} \right) \; .$$

But the projection of $\vec{x}$ onto $\vec{v}$ is

$$\vec{x}_{\parallel} = \frac{(\vec{x} \cdot \vec{v}) \vec{v}}{v^2}$$

hence

$$\vec{x}_{\perp} = \vec{x} - \frac{(\vec{x} \cdot \vec{v}) \vec{v}}{v^2}$$

Substituting this into $\vec{x}' = \vec{x}_{\shortparallel}' + \vec{x}_{\perp}'$ and using the preceding transformation formulae, one obtains the Lorentz transformation for the situation when the relative velocity $\vec{v}$ has an arbitrary direction:

$$\vec{x}' = \vec{x} + (\gamma - 1)\,\frac{(\vec{x}\cdot\vec{v})\,\vec{v}}{v^2} - \gamma\,\vec{v}\,t$$

$$t' = \gamma\left(t - \frac{(\vec{x}\cdot\vec{v})}{c^2}\right)\ .$$

(2.8)

One should observe that, since $\vec{x}_{\perp}\cdot\vec{v}=0$, the vector $\vec{x}_{\shortparallel}$ in the inner product $\vec{x}_{\shortparallel}\cdot\vec{v}$ may be replaced by $\vec{x}$.

2.4.4 Interval and the Principle of Invariance

The special Lorentz transformations that we derived in the foregoing and which depend on the three components of $\vec{v}$, do not yet yield the complete Lorentz group. The spatial rotations must also be considered, which adds three more parameters. In view of the introduction of the four-dimensional Minkowski space (see later), we ought to note at this point that the general Lorentz transformation can be characterized by means of a principle of invariance. This is done in analogy to the rotation group, where we know that it consists of the transformations which leave the distance of the point $\vec{x}=\{x,y,z\}$ from the origin of the coordinate system unchanged. The norm of the vector $\vec{x}$, i.e. also $\vec{x}^2$, is invariant under rotations.

Under the Lorentz transformations (2.8), time is also transformed. An event in system K is specified by stating the time and location $\{t,x\}$ of it. It is plausible (and it turns out to be advantageous) to introduce in a corresponding four-dimensional space a coordinate system where the three space coordinates and the time are used to label the axes. In this space, the coordinates of events are represented by the "world points" $\{ct=x^0,\vec{x}\}$. Taking $x^0=ct$ for the fourth coordinate, all coordinates will have the dimension of length. In this space, there corresponds to the motion of a particle a line called the world line of the particle, which, in general, is curved.

For a particle with a uniform and rectilinear motion the world line is a straight line.

In order to find the invariant of the Lorentz transformations which corresponds to the norm of a vector $\vec{x}$ in the case of rotations, the following brief consideration proves useful. We start with the preceding result which was that light propagation occurs with the limiting speed $\sigma = c$, independently of the reference frame. We imagine that at $t = t' = 0$, when the two inertial systems K and K' coincide, a light signal is emitted from the common point of origin. The light signal propagates like the surface of a sphere with increasing radius; this sphere is described in K and K', resp., by the equation

$$c^2 t^2 - \vec{x}^2 = 0 \qquad \text{or} \qquad c^2 t'^2 - \vec{x}'^2 = 0 \ . \tag{2.9}$$

Here we utilized the fact that the light velocity c is independent of the inertial system. The quantity

$$s = \left[c^2 t^2 - \vec{x}^2 \right]^{1/2} \tag{2.10}$$

is called the separation (or better, the interval) between the two world points $\{ct, \vec{x}\}$ and $\{ct = 0, \ \vec{x} = 0\}$. According to equations (2.9), vanishing of the interval between two world points connected by a light signal is an invariant statement. If the interval between two events vanishes in one system, then it does also in all other systems. But when passing from K to K' the same Lorentz transformation (2.6) holds for all world points, whether or not they are connected by a light signal. Therefore one expects that also non-zero intervals between arbitrary world points are invariant. A brief calculation, left to the reader, convinces one that, in general,

$$s^2 = c^2 t^2 - \vec{x}^2 = c^2 t'^2 - \vec{x}'^2 \tag{2.11}$$

so that s^2 is invariant against the special Lorentz transformation (2.6). We emphasize that this invariant quantity may have not only the value zero but it can have also any nonvanishing positive or negative

36

value. Since the choice of the origin is arbitrary, one has, more generally,

$$s_{12}^2 = c^2(t_2 - t_1)^2 - (\vec{x}_2 - \vec{x}_1)^2 \qquad (2.12)$$

representing the square of the invariant separation between the world points $\{ct_2, \vec{x}_2\}$ and $\{ct_1, \vec{x}_1\}$. When two events are infinitesimally near, then the corresponding invariant quantity, the squared four-dimensional line element ds, is given by

$$ds^2 = c^2 dt^2 - d\vec{x}^2 \qquad (2.13)$$

Thus, in analogy to the spatial rotations, we may characterize the general Lorentz transformations by an invariance property. The in-homogeneous Lorentz group, usually called the Poincaré group, consists of all transformations between inertial systems that leave the squared separation s_{12} of events, defined by (2.12), invariant. This group has ten parameters, because, besides the special Lorentz transformations and spatial rotations it also contains the translations of space and time.[1] In addition, spatial and temporal reflections belong to the group as well. Later on we shall discuss the Lorentz group in more detail.

2.5   Kinematical Consequences of the Lorentz Transformation

Before tackling, in Chapter 5, relativistic dynamics, we wish to discuss the most important kinematic consequences of the Lorentz transformations; because of their unusual nature, these often lead to difficulties in understanding. The so-called paradoxes belong to this category; the cause of these apparently contradictory statements may be traced to an inadequate consideration of the concept of rela-tivistic simultaneity.

----

[1] We recommend the reader to derive the special Lorentz transfor-mations from this invariance principle.

2.5.1 Contraction of Length

It follows from the Lorentz transformation (2.6) that the length of a moving meter stick is reduced by a factor $\sqrt{1-\beta^2}$ in the direction of motion.

Let a meter stick rest in K', extending along the x' axis. Suppose a measurement in K' determines that the endpoints of the stick lie at x_2' and x_1'. Thus, its length at rest is

$$l_0 = x_2' - x_1'$$

What will be the length of the stick for an observer in the inertial system K relative to whom the meter stick moves along the x axis with the velocity v? This question may be answered by calculating the coordinates as changed under the Lorentz transformation (2.6). Let the coordinates of the endpoints of the meter stick, as measured by an observer in K at time t be x_1 and x_2. Consequently, its length as determined in K is $l=x_2-x_1$. Substituting these coordinates into the Lorentz transformation and using $t_1=t_2=t$, one obtains

$$x_2' = \gamma(x_2 - vt) \quad , \quad x_1' = \gamma(x_1 - vt)$$

Upon taking the difference, the time dependent terms cancel and one gets

$$x_2' - x_1' = \gamma(x_2 - x_1) \quad ,$$

i.e., with $l_0=l'=x_2'-x_1'$ and using the definition of γ we have

$$l = l_0\sqrt{1-\beta^2} \quad , \quad l'= l_0 \quad . \tag{2.14}$$

The annotation $l'=l_0$ is ment to make it clear that the length at rest is measured here in the system K'. Thus, the length of the meter stick in relative motion to K, as measured in the system K, is shorter by a factor $\sqrt{1-\beta^2}$ then its proper length (length at rest) l_o. [1]

Since the dimensions of the meter stick in directions perpendicular to the velocity are not altered, the relation between the volume measured in K and the proper volume V_o is given by

$$V = V_0 \sqrt{1 - \beta^2} \tag{2.15}$$

Thus, length and volume are relative. Herein rests a fundamental difference versus classical physics, where these concepts are taken to have absolute meaning. For sufficiently small velocities, $v^2 \ll c^2$, one may neglect the contraction of length.

The reader must be reminded that the two reference systems are fully equivalent. Therefore, if the meter stick is at rest in K, then its length measured in K' will be again shorter by the above given factor than its proper length in K. Of course, this reciprocity property of length contraction can also be seen directly from the Lorentz transformation. In this regard, the relativistic interchange can be a useful consideration. When deriving $l = l' \sqrt{1-\beta^2}$ (where l' is to be identified with the proper length l_o), we used the transformation from K to K'. If now the meter stick is at rest in K, then one must set l equal to the proper length l_o, ($l = l_o$), and the calculation of the change goes from K' to K. Thus, from the previous result (2.14), i.e. $l = l' \sqrt{1-\beta^2}$, $l' = l_o$, it follows by means of relativistic interchange and identification of the length at rest that

$$l' = l_0 \sqrt{1 - \beta^2} \ , \qquad l = l_0 \tag{2.16}$$

Equations (2.14) and (2.16) do not contradict each other, because

---

[1] Expressions such as saying that "the meter stick appears shortened when looking at it from the system K" may lead to misunderstanding if they are not more closely specified. The measurement of length (or more generally any observation) in K implies only the determination of event coordinates. The rash opinion that under simple conditions the contraction of length would be directly visible (as perhaps on a photograph) was wide spread, and its incorrectness was pointed out only late (1959). We shall return to this point.

the quantities which occur in them pertain to different situations. In one of the situations the pair $\{t_1,x_1; \ t_2,x_2\}$ of events is simultaneous in K, but in the other situation the pair of events $\{t_1',x_1'; \ t_2',x_2'\}$ is simultaneous in K'. However, the concept of simultaneity no longer has an absolute meaning, but it makes sense only if the relevant reference system is specified. This holds true also for the length of a meter stick, which obviously is measured by simultaneous marking of the positions of its endpoints. Thus, the lengths 1 and 1' in equations (2.14) and (2.16) have a different meaning, and in each case the stationary observer finds a contraction of the meter stick which is in relative motion to his system.[1]

The question, "which of the two lengths is really shortened", has no meaning in this context. It would be equivalent to the question "which of the two inertial systems K and K' is at absolute rest". According to the principle of relativity one cannot find an inertial system at absolute rest, and the statements discussed above hold true with respect to the relevant point of view, i.e., the reference system used by the observer.

One must be aware of the fact that the length contraction (2.14), in contrast to the contraction suggested by Lorentz, is not caused by forces which deform the body. The relativistic length contraction is a kinematical effect and expresses the compatibility of measuring space and time on one hand and the principle of relativity, on the other.

2.5.2 Dilation of Time

It follows from the Lorentz transformation that time intervals measured by a moving clock are dilated by a factor $(1-\beta^2)^{-1/2}$. In other words: moving clocks go slower than clocks at rest.

Take a clock at rest, located at x' in system K', moving the relative velocity v along the x axis of the stationary system K. Let us now consider two events that occur at the location x' at

---

[1] A useful analog, taken from everyday life is the following: If two people move away from each other, to each one it appears that the other one is getting smaller.

two subsequent instants t_1' and t_2'; take, for example subsequent pointer indications of a clock at x'. This time, as determined in system K' by a clock at rest in K', is called the "proper time" t_o. In consequence of the Lorentz transformations, an observer in K will find, using his clock stationary in K, the following times for these events:

$$ t_2 = \gamma \left(t_2' + \frac{v}{c^2} x' \right) \ , \quad t_1 = \gamma \left(t_1' + \frac{v}{c^2} x' \right) \ , \quad x_2' = x_1' = x' $$

Thus, the time lapse between the events in question, as seen in K, is

$$ t_2 - t_1 = \gamma \left(t_2' - t_1' \right) \ , $$

i.e., writing $\Delta t_o \equiv \Delta t' = t_2' - t_1'$, and $\Delta t = t_2 - t_1$, we have

$$ \Delta t = \frac{\Delta t_o}{\sqrt{1 - \beta^2}} \ , \quad \Delta t' = \Delta t_o \qquad (2.17) $$

The way of writing "$\Delta t' = \Delta t_o$" is to remind us that the proper time interval Δt_o is measured in System K'. As seen from equation (2.17), the time interval shown by a moving clock and measured in the stationary system K, is longer by a factor $(1 - \beta^2)^{-1/2}$ than the proper time interval Δt_o. Thus, the proper time interval of a moving clock is always shorter than the corresponding time lapse in the stationary system. In other words: moving clocks go slow.

Since the systems K and K' are equivalent, time dilation is also a reciprocal effect. For a clock at rest in K at $x_1 = x_2$, we obtain from (2.17), with a relativistic interchange and identification of the proper time interval as $\Delta t = \Delta t_o$, the following relation:

$$ \Delta t' = \frac{\Delta t_o}{\sqrt{1 - \beta^2}} \ , \quad \Delta t = \Delta t_o \qquad (2.18) $$

This equation says that a clock at rest in K goes slow for the observer in K'.

Thus, for any stationary observer, clocks in a system which is in relative motion appear to go slow. This fact is in accord with the relativity principle. We do not have a contradiction (there is no "clock paradox"), because the pairs of events described in the equations above, relate to different situations. In one situation, the two events occur in K' at the same location $(x_1'=x_2')$, and in the other situation, they occur in K at the same location $(x_1=x_2)$. And the statement that two events occur at the same location is meaningful only if one specifies the relevant frame of reference. Hence, the time intervals in (2.17) and (2.18) have different meanings. In one case one has to identify $\Delta t'$ with the proper time interval, in the other case, Δt. The question "which clock is really slow", is just as meaningless as "which inertial system is really at rest". The statements of both the equivalent observers, referring to their respective frame of reference, are both correct.

There are numerous experimental verifications of time dilation. A particularly instructive case is the decay of π mesons (pions) in the laboratory frame. These particles are produced by bombarding a suitable target in an accelerator with high energy protons. They emerge from the target with a velocity near to light velocity. The unstable π mesons decay into muons with appropriate charge ($\mu^\pm$) and neutrinos (ν_μ),

$$\pi^\pm \longrightarrow \mu^\pm + \nu_\mu$$

The half-life of a pion at rest is $t_{1/2}=1.8 \times 10^{-8}$ sec, i.e., half of the pions present at a given time has decayed after 1.8×10^{-8} sec.[1] In the laboratory one finds that the intensity of a beam of pions with the velocity 0.99 c falls to its half value at a distance of about 38 m from the target. Using $t_{1/2}=1.8\times10^{-8}$ sec and $v=2.97\times10^8$ m/sec (=0.99c), one can calculate the distance d within which half of the pions decayed. Not considering time dilation, one would find d=5.35 m and not the measured value 38 m. However, this calculation is not correct, because $t_{1/2}$ is measured in the rest system of the pion,

---

[1] In contrast to the half-life $t_{1/2}$, the mean life τ is defined as the reciprocal of the decay constant λ which occurs in the decay law, i.e. $\tau = 1/\lambda$. From the two definitions it follows that $t_{1/2} = \tau \ln 2 = 0.693 \ \tau$. Tables usually quote τ.

but the distance d is measured in the laboratory system. Relative to the stationary lab system, the pion moves with a high velocity. Therefore, one must take into account the time dilation. In the lab system, one measures a dilated time interval for the half-life:

$$t_{1/2} = \frac{t_{1/2}^0}{\sqrt{1 - \beta^2}} = \frac{1.8 \times 10^{-8} \text{ s}}{\sqrt{1 - (0.99)^2}}$$

The numerical value $t_{1/2} = 12.85 \times 10^{-8}$ sec obtained from this equation leads to the longer distance $d = vt_{1/2} = 38.2$ m, in agreement with experiment.

The situation may also be illuminated via the length contraction effect. So far the measurements of distance and time were done in terms of the lab system K. Now we switch to the System K' attached to the pion and ask, what is the distance for which the pion beam's intensity falls to half its value. As seen from the pion, the laboratory moves with near light velocity (0.99 c), which implies that the 32.2 m observed therein appear shortened in K', viz.

$$d' = 32.2 \sqrt{1 - (0.99)^2} \text{ m} = 5.3 \text{ m}$$

Now, the time needed for pions with velocity v=0.99 c to traverse this distance, is equal to half-life,

$$t_{1/2}^0 = \frac{d \sqrt{1 - \beta^2}}{v}$$

The value $t_{1/2} = 1.8 \times 10^{-8}$ sec which follows from this, equals the experimental half-life quoted above. This example illuminates the physical reality of time dilation and length contraction, which follow from the Lorentz transformation. With precise measurements of the pion and muon lifetimes, the time dilation effect was verified to an accuracy of 0.4% and 0.1%, respectively.[1]

---

[1] D.S. Ayres et al., Phys. Rev. D3, 1051(1971); J. Bailey et al., Nature 268, 301(1971).

In general, we can say that, sofar, all measurements in high-energy physics agree with predictions from time dilation and length contraction. In fact, all experiments and even the accelerators themselves are so designed that they take into account the relativistic effects. Thus, application of special relativity to the area of high-energy physics, including the construction of accelerators and the relevant technology, became a routine affair.

Interestingly, in 1971 J. Hafele and R. Keating[1] provided also a direct determination of time dilation, using macroscopic clocks. These authors recognized that the accuracy of a few nanoseconds (10^{-9} sec) per day, possessed by transportable atomic clocks based on ^{133}Cs, should be sufficient to measure the influence of velocity and gravitational field on the rate of clocks which are carried around in simple commercial aeroplanes. With such velocities, one can expect, after one orbit around the Earth, a time difference of 10^{-7} sec relative to a clock at rest on Earth. When the experiment was performed, the Cs clocks were flown eastward and westward around the Earth. In this way, it was possible to separate off the effect of gravity on the clock rate[2] from the velocity dependent effects of present interest. In such a manner, the time dilation given by (2.17) could be verified, with relatively simple means, with an accuracy of about 10%.

2.5.3 Non-Synchronism of Moving Clocks

A further consequence of the Lorentz transformation is that, for a stationary observer in K, the readings of moving synchronous clocks in K', are not synchronous. One can see this from the equation

$$ t = \gamma \left(t' + \frac{v}{c^2} x' \right) $$

---

[1] J.C. Hafele and R.E. Keating, Science 177, 166 and 168(1972).

[2] The clock in the aeroplane at some hight above the surface of the Earth is exposed to a slightly weaker gravitational potential, and therefore it goes faster than the clock of comparison on the ground.

Let us choose a fixed instant t in K. Then the right-hand side of this equation has a fixed value; but then, the bigger x' the smaller must be t'. Thus, the farther is a clock in K' on the x' axis, the more will its time reading be late. Hence, for the observer in K the clocks in the moving system are no longer synchronous. This statement is reciprocal, and it is but an other expression of the fact that the concept of simultaneity depends on the system of reference.

2.5.4 Velocity Transformation

Let the system K' move relative to K with the velocity w along the positive x axis. Let $\vec{v}=d\vec{x}/dt$ be the velocity of a particle in system K, and $\vec{v}'=d\vec{x}'/dt'$ its velocity in K'. The Lorentz transformation for the differentials,

$$dx = \gamma \, (dx' + w \, dt')$$
$$dy = dy'$$
$$dz = dz'$$
$$dt = \gamma \, (dt' + \frac{w}{c^2} \, dx')$$

yields, if one divides the first three equations by the fourth and uses the above definitions of $\vec{v}$ and $\vec{v}'$:

$$v_x = \frac{v_x' + w}{1 + \frac{v_x' w}{c^2}} \quad , \quad v_y = \frac{v_y' \sqrt{1 - \beta^2}}{1 + \frac{v_x' w}{c^2}} \quad , \quad v_z = \frac{v_z' \sqrt{1 - \beta^2}}{1 + \frac{v_x' w}{c^2}} \qquad (2.19)$$

Similarly, for $\vec{w}$ in an arbitrary direction we get from (2.8) the more general relation

$$\vec{v} = \frac{1}{1 - \frac{\vec{v} \cdot \vec{w}}{c^2}} \left\{ \sqrt{1 - \beta^2} \, \vec{v} + \left[(1 - \sqrt{1 - \beta^2}) \, \frac{\vec{v} \cdot \vec{w}}{w^2} - 1 \right] \vec{w} \right\} \qquad (2.20)$$

One sees that the velocities $\vec{v}$ and $\vec{w}$ do not play a symmetrical role in this equation. The reason is that, in general, Lorentz transformations do not commute. For $w_x=w$, $w_y=w_z=0$, $v' \parallel w$ one obtains the already familiar (cf.equ.(2.7)) addition law for parallel velocities,

$$v = \frac{v' + w}{1 + \frac{v'w}{c^2}} \qquad (2.21)$$

How will this change the angle spanned between the velocity $\vec{v}$ and, say, the x axis? Let us choose the coordinate system so that $\vec{v}$ lies in the xy plane. Then the velocity components in system K and K', respectively, are given by

$$v_x = v \cos\theta \,, \quad v_y = v \sin\theta$$
$$v_x' = v' \cos\theta' \,, \quad v_y' = v' \sin\theta' \,,$$

where $\theta(\theta')$ denotes the angle between $\vec{v}(\vec{v}')$ and $x(x')$ axis in system $K(K')$. With this, the transformation equation (2.19) gives

$$\frac{v_y}{v_x} = \tan\theta = \frac{v' \sqrt{1-\beta^2} \, \sin\theta'}{v' \cos\theta' + w} \qquad (2.22)$$

With this formula one can calculate the directional change of velocity if one passes to another inertial system.

In the special case of directional change of light when passing to another reference frame, one must set $v=v'=c$; in this manner one obtains the relativistic expression for light aberration:

$$\tan\theta = \frac{\sin\theta' \sqrt{1 - \frac{w^2}{c^2}}}{\cos\theta' + \frac{w}{c}} \qquad (2.23)$$

From the formula (2.19), similar expressions for $\sin\theta$ and $\cos\theta$ may be deduced.

An interesting application of the velocity addition law is the interpretation of the Fizeau drag-experiment. In a medium with refractive index n, a light signal propagates with velocity $v'=c/n$, provided one disregards dispersion. If now this medium streams along with a velocity w parallel to v', then an observer at rest will find, if he uses the addition law, that the light velocity is

$$c_M = \frac{\frac{c}{n} + w}{1 + \frac{w}{nc}}$$

i.e., in the approximation for $w \ll c$,

$$c_M \approx \left(\frac{c}{n} + w \right)\left(1 - \frac{w}{nc} \right)$$

and if one neglects w^2/nc, then one has

$$c_M \approx \frac{c}{n} + w \left(1 - \frac{1}{n^2} \right) . \tag{2.24}$$

The so-called Fresnel drag coefficient $(1-1/n^2)$ describes the deviation from the classical velocity addition law. This result is verified by the many variants of the Fizeau experiment.

Let us emphasize at this point that the parameter w in the Lorentz transformation (2.20) denotes the velocity with which realistic bodies (and the connected reference systems) move. The limiting velocity c, which cannot be reached by such bodies, is the maximal propagation velocity of effects. In contrast, one can easily think of velocities greater than c, but which have only a kinematical or geometrical character and which do not correspond to the motion of realistic bodies or to the propagation of interactions (signals).

For example, if a light ray rotates with a sufficiently high angular velocity ω, one can achieve that the light spot on a surface at a distance R moves with the velocity $v = \omega R$ which is greater than the velocity of light. This does not contradict relativity theory, because it is not possible to transmit some energy or a signal between two locations on the surface in this way. To do so, one ought to switch on-and-off (or modulate) a light ray that directly connects the two locations. But the propagation velocity of this signal is (in vacuum) the light velocity c.

In optical media which, for a certain wavelength domain have a refractive index $n < 1$, one gets phase velocities exceeding c. However, since there is no modulation and since the monochromatic wave train has an infinite extension, one cannot transmit signals with the phase velocity. In contrast, the propagation of discontinuities of the electromagnetic wave (switching on and off) occur with a front velocity v_F which equals the light velocity c.[1]

Because of the existence of the maximal velocity c for propagation of effects, the concept of an ideal rigid body is incompatible with the theory of special relativity.[2] An ideal rigid body, i.e. one with unchanging distances between its constituent parts, could be used for transmitting a signal with a practically infinite velocity. A knock on one end would instantaneously be detectable at the other end. For deformable bodies, in contrast, the caused deformation propagates as a shock wave in the body with velocity v < c.

2.5.5 The Meter Stick Paradox

If one fails to take well into account the relative meaning of the concepts one uses, it is possible to formulate in the theory of special relativity a bunch of seemingly contradicting statements. The resolution of these so-called paradoxes can be reduced, in the ultimate analysis, to the relativity of simultaneity. In the examples given below, we will first of all use the length contraction effect to formulate such "paradoxical" statements.

Suppose a 10 cm long rod moves along the x axis, parallel to a tabletop which has a 10 cm long hole in it (see Fig. 2.3). Suppose the tabletop with the hole moves upward the y axis with velocity u, so that (as seen from the reference system K of the table) at t=0 the midpoint of the rod coincides with the midpoint of the hole. If the velocity v_s of the rod is so high that

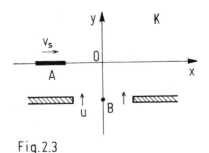

Fig. 2.3

---

[1] A detailed discussion of signal propagation (group and front velocity) may be found in L. Brillouin, Wave Propagation and Group Velocity (Academic Press, New York 1960).

[2] This also applies to "incompressible" fluids.

the factor of length contraction is 1/2, then, because of the length contraction, the rod measures only 5 cm in the rest system K of the table, and it can pass through the hole. But, as seen from the rest system K' of the rod, the hole is shortened to 5 cm. So, can the rod pass through or can it not? The contradiction rests not in the circumstance that the distances are mutually forshortened, but rather in the statement regarding passing or not passing through. It should be possible to make about this an unambiguous statement independently of the reference frame. Actually, the consideration of the fact that in the system K' (rest frame of the rod) the rod and the tabletop are no longer parallel but are inclined to each other, is crucial.

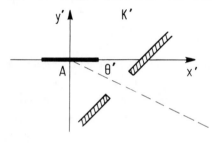

Fig.2.4

Also, the center of the hole B no longer approaches the center of the rod A perpendicularly to the x' axis, but rather along a straight line which makes an angle less than 90° to the x' axis (see Fig.2.4). Using (2.2), it is easy to calculate this angle. With a relativistic interchange one has

$$\tan \theta' = \frac{v \sqrt{1-\beta^2} \sin \theta}{v \cos \theta - w}$$

In our example, one must take $w=v_s$, $v=u$, and $\Theta = 90^\circ$. So we get

$$\tan \theta' = - \frac{u\sqrt{1-\beta^2}}{v_s},$$

i.e. the point B moves along the straight line $y' = -ux'/(v_s \gamma)$. If the slope $-u/(v_s \gamma)$ is moderate, the angle between this straight line and the x' axis is small and the rod, even as seen from system K', will pass through the hole.

A further example is the following, apparently paradoxical problem. A person carrying horizontally an 8 m long pole, approaches an open garage of 4 m length with the velocity $v=(\sqrt{3}/2)c=0.866$. For an observer in the system K of the garage at rest, the pole appears only 4 m long on account of the length contaction ($\gamma=2$). Thus, he can close the trapdoor in front when the forward end of the pole reaches the backwall of the garage (made of concrete). So, in system K, it is possible to carry the pole into the garage. The pole, brought to rest, will try to assume its original length relative to the garage. Provided it survived the shock, it must bend or brake open the garage door.

On the other hand, from the viewpoint of the person who approaches the garage door with velocity v, the garage appears only 2 m long. Should he therefore worry that he will lose 6 m of his pole when the garage door shuts close? According to the relativity principle, this catastrophe does not occur. The statement that the pole fits into the garage, cannot depend on the choice of the inertial system. One must be aware that signals propagate with a finite velocity and the concept of simultaneity is no longer an absolute one. While in the system K in rest relative to the garage the appearence of the pole at the end of the garage and the shutting of the door occur simultaneously, these events are no longer simultaneous in the rest system K' of the person.

The sequence of events in system K' can be described in the following way. The open garage moves toward the pole with the high velocity v. It keeps moving when the tip of the pole hits the concrete wall and it takes along the forward end of the pole. Because of the finite propagation velocity of effects, the far end of the pole is still at rest. The signal caused by the impact of the pole propagates backward along the pole by means of a shock wave. In order to reach the far end, the shock wave must pass a distance of 8 m, but the door front has to do only 6 m. Even if the shock wave propagated with light velocity, the race between the shock wave and the door front toward the far end of the pole will end in a draw. The condition for this to be the case is, on account of the distances passed, that $v=(6/8)c=$ 0.75c. But in our example $v=0.866c > 0.75c$. This implies that the door front of the garage reaches the far end of the pole at rest before the shock wave, and one can close the door behind the pole. As one would expect from the principle of relativity, no catastrophe occurs in the system K'.

2.5.6 The Twin Paradox

It seems that this paradox caused most of the controversies and it has been discussed repeatedly in numerous papers.[1]

A space traveller B, who took a trip with a velocity v comparable to the light velocity, will find upon his return to Earth that, because of the time dilation, he aged less than his twin brother A who stayed on Earth. Whosoever travels, stays younger. If one uses the relativity principle carelessly, one is easily led to a paradoxical situation. Because of the "relativity" of viewpoints, each of the twins could consider the other twin as the traveller. Consequently, each of them should find upon their return the other one less aged – which is a contradiction.

However, this kind of reasoning is based on ignoring the assumptions of the theory of special relativity. The reference frames in question are not equivalent in the sense of the relativity principle. The twin who stays at home, is permanently in an inertial system. In contrast, because of the necessity of periods of acceleration and deceleration, the space traveller is not in an inertial system (at least not at all times). Even though one can choose the acceleration periods arbitrarily small and eliminate the acceleration effects

---

[1] See, for example, L. Marder, Time and the Space Traveller (University of Pennsylvania Press, Philadelphia, 1971) where a detailed list of references is given, containing 241 entries.

on the clocks,[1] nevertheless the space traveller must alter his reference frame when he returns. Hence, there is really no symmetrical, i.e. interchangeable, situation between the two twins. Consequently we do not have a paradox, and the asymmetry between the twins' ages is as real as is the time dilation found for decaying pions, or the time dilation experimentally observed with moving clocks by Hafele and Keating (see p.43).

    Having made this preliminary commentary, let us clarify in detail the dependence of the proper time on the path; this is the crux of the matter. Instead of twins, let us consider two clocks, which in a common rest system show the same proper time. If, after passing through two different routes, they are again brought to coincide, their readings will no longer agree. The time shown by a moving clock depends on the route taken. Figure 2.5 shows the routes (world lines 1 and 2) of the resting clock (A) and the moving clock (B), represented in the rest system (ct,x) of A. According to equation (2.17), the relation between the proper time differential $d\tau$ (given by $d\tau \equiv dt_o$) and the time differential dt referred to the system of A, is given by

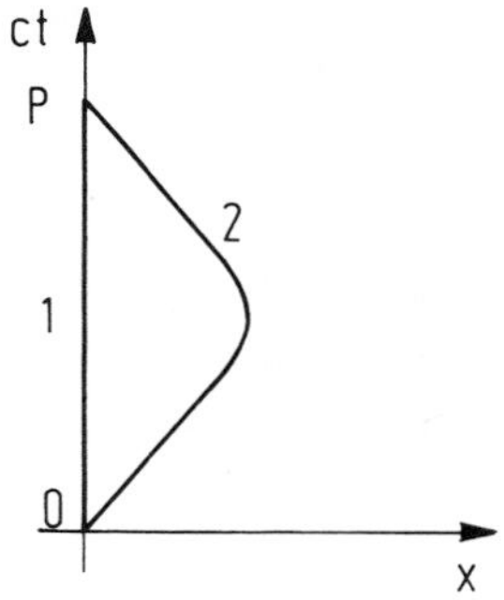

Fig. 2.5

$$d\tau = dt \sqrt{1 - \beta^2}$$

(2.25)

---

[1] This is achieved by using an ideal clock which is insensitive to accelerations. Such a property depends, of course, on the construction of the clock. - Thus, for example, a pendulum clock would surely not be suitable, since it consists of the clock proper including the clockcase and the whole Earth. In contrast muons, having a decay time unchanged even by strong external forces, could serve as ideal clocks, in a very good approximation. - In order to justify this "clock hypothesis", we note that in principle it is possible to construct an ideal clock, because accelerations have an absolute meaning in our context. One could therefore equip the clock with an accelerometer and a servo mechanism which corrects the clockreadings in accord with the value of the determined acceleration. In contrast, the velocity dependent effect (2.17) cannot be eliminated.

Consequently, the proper time interval between the events 0 and P is the sum (i.e., in this case, the integral)[1]

$$\Delta\tau = \int_0^P dt \sqrt{1 - \frac{v^2(t)}{c^2}}$$
$$= \int_0^P \sqrt{dt^2 - \frac{dx^2}{c^2}} = \frac{1}{c} \int_0^P ds$$

(2.26)

where the integration path must be taken along the relevant world line.

Let us now consider path 1 followed by the clock A, which is at rest on the x axis. Since dx=0, the proper time that passes while traversing this route is

$$\Delta\tau_1 = \int_{(1)} dt = t_P - t_0$$

This time difference will be shown by the clock at rest. On the other hand, the time that passed if path 2 is followed, is (as measured in the system of A)

$$\Delta\tau_2 = \int_{(2)} dt \sqrt{1 - \frac{v^2(t)}{c^2}} = \int_{(2)} \sqrt{dt^2 - \frac{dx^2}{c^2}}$$

Since now dx≠0, $\Delta\tau_2$ and $\Delta\tau_1$ will be different. If v(t) were known, we could evaluate the integral. For an estimate, we do not really have to know v(t). Since dx^2 is always positive, one has

---

[1] We assume that along the world line 2 a clock is used which is insensitive to accelerations. The rate of such an accelerated ideal clock is at every instant t the same as the rate this clock would have if it were in uniform motion (i.e. in an inertial system) with the same velocity v(t). Therefore relation (2.25) holds at every instant t. - The above procedure is analogous to the well known geometrical rectification method employed for the determination of the length of a curve from the sum of sections which approximate it.

$$\Delta \tau_2 < \Delta \tau_1$$

When the clocks meet again at P, they will show different times, with the moving clock being late compared with the resting clock. – The travelling twin stays younger.

The system (ct,x) connected with A is an inertial system. The motion of the clock of traveller B is described in this system by a curved world line, since this clock is subject to accelerations and does not move on a path with a uniform velocity. Otherwise the world line would be a straight line with a slope $> 45^{\circ}$, and there would be no return to the stationary clock.

Special relativity can predict the behaviour of accelerated objects as long as one uses an inertial system for the formulation of physical laws. The above given calculation of the invariant proper times $\Delta \tau_1$ and $\Delta \tau_2$ along the different paths provides an example. In contrast, a frame of references connected to the space traveller B would not be an inertial system. One can very well convert to the accelerated (i.e. curvilinear) coordinate system and determine the correct proper time in this system. The corresponding mathematical formalism is usually developed only in the context of general relativity. But for the explanation of the twin paradox it is not necessary to resort to this somewhat complicated conversion.[1] For the problem on hand, i.e. the calculation of the path dependent proper time, one can rely throughout on an inertial system (the rest system of A) and in this way one arrives at the correct result in a simpler manner. We reiterate: the situation of the twins is not symmetrical as is erroneously assumed when the paradox is formulated. The twin who remains at home stays all the time in a single inertial frame, the other does not.

---

[1] A general discussion of physical laws in accelerated systems can be found in H. Heintzmann and P. Mittelstaedt, Springer Tracts in Modern Physics 47, 185 (1968). See also the more recent paper by E.A. Desloge and R.J. Philpott, Am. J. Phys. 55, 252 (1987).

It is possible to explicitly calculate $\Delta\tau_2$ for an idealized variant of the travel of a twin (of a clock). We assume that the space traveller moves along with a constant velocity v except at the start, at the turning-round, and at the end of the trip. The contributions of these acceleration phases to $\Delta\tau_2$ should be taken negligibly smaller than the contributions from the states with constant velocity. Such a situation may be thought of as realized in the following manner. Imagine the travel, divided into two parts, such that we have two spaceships which meet and exchange signals, where one spaceship moves away with a uniform speed from the stationary system A (start in flight), whereas the other spaceship approaches (also with uniform velocity) the system A. (This set-up is visualized in Fig.2.6.) Let the point of intersection of these straight spaceship world lines (point Q) be associated with $t = T/2$. (For brevity, we use the notation $T = t_p - t_o$.) To start with, we get for the integral along the direct route OP the value

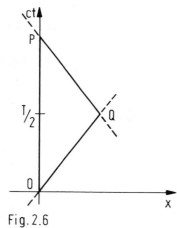

Fig. 2.6

$$\Delta\tau_1 = \int_0^P dt = T .$$

For the partial sector OQ we have dx=v dt, v=const., so that

$$\int_0^Q \sqrt{1 - \beta^2}\, dt = \sqrt{1 - \beta^2}\, \frac{T}{2} .$$

Finally, along QP we have dx=-v dt, i.e.

$$\int_Q^P \sqrt{1 - \beta^2}\, dt = \sqrt{1 - \beta^2}\, \frac{T}{2} .$$

Putting everything together we then obtain the shortening of the proper time $\Delta\tau_2$ vis-a-vis $\Delta\tau_1$. As expected for a constant velocity from equation (2.17) we indeed have

$$\Delta \tau_2 = T \sqrt{1 - \beta^2} < T = \Delta \tau_1 \ .$$

The asymmetry is here also manifest. Even if the traveller sojourns predominantly in an inertial system, he must switch from one inertial system (outward trip) to another one (return trip). In contrast, the twin staying home, does not change his inertial frame.

This observation suggests a comparison with the situation regarding ordinary curves in the plane which one obtains from Fig.2.5 when one replaces there ct by a spatial coordinate.[1] Then it is immediately obvious that the curved path 2 is longer than the straight connection between O and P. In addition, it will be clear that while the greater length is caused by the curving (acceleration) at the point of turning around, substantial contributions come only from the rectilinear (unaccelerated) sections. In exact analogy, the clock brought back by the travelling twin is retarded not because it has been exposed during a certain time to accelerations, but rather because it has been accelerated at all; or putting it in another way, because there is a change of the inertial system at return.

Should one think that the space traveller B is "really" younger (in the biological sense of the word) than his twin brother A who stayed at home? So far there is no hint regarding a difference between physics for organic processes and physics for the inorganic matter involved in these processes. Therefore one must expect that the biological functions of B are slowed down in the same measure as his physical clock. Here again an anology may prove useful. It is generally accepted that, by diminishing the temperature, life processes may be slowed down. Just like temperature differences are real and measurable, so also the different states of motion of the twins, i.e. the asymmetry of their reference frames, is real and can be determined.

---

[1] This comparison was given by H. Bondi, Assumption and Myth in Physical Theory (Cambridge Univ. Press, Cambridge 1967).

In view of the frequent fantastic speculations about barely
aging space travellers, as one finds in the popular-science and enter-
taining literature, the reader should bear in mind that the effect
can be noticed only for travel speeds which are near to c.

2.5.7 Observation of Moving Objects

Now we turn to the question: is it possible to directly see
the length contraction? As already observed in the footnote on p.38,
one must not misinterpret the notion of an observer in the special
theory of relativity. In general, the term signifies merely the determination of coordinates in an inertial system. If one has in mind
"true" observation or "seeing", then one must take into account the
finite time of passage of a light signal from the object to the observer. Since this was not done properly, for a long time it was thought
widely that the length contraction should be visible. That this is
an improper representation of length contraction became evident
only much later.[1]

The time spans which are needed to get from the diverse parts
of a moving object to the eye of a stationary observer for to arrive
there simultaneously, are different, and therefore correspond to different locations of the body. It then follows that, in general, the
body will have a distorted appearance. If the moving body is sufficiently far away so that the light rays reaching the observer are
essentially parallel, the body will appear as though slightly rotated.
We illuminate this with the following simple example.

A rectangular parallelepiped with length $L_o(\overline{AD})$, breadth $B_o(\overline{AF})$,
and height $H_o(\overline{AB})$ moves parallel to the edge AD. We observe it from
a direction perpendicular to the surface ABCD and hence perpendicular
to the velocity v (see Fig.2.7a). Suppose the observer is so far away
that the light rays originating at the parallelepiped can be taken

---

[1] R. Penrose, Proc.Cambr.Phil.Soc.55, 137 (1959); J. Terrell, Phys.
Rev. 116, 1041 (1959). See also V.F. Weisskopf, Physics Today,
Sept. 1960, p. 24; G.D. Scott a.M.R. Viner, Am. J. Phys. 33, 534
(1965). - The much earlier work on this topic by A. Lampa, Z. Physik
72, 130 (1924), has largely been unrecognized.

to be parallel. Light coming from the rear edge FE must traverse a path which is longer by B_o than the path from the edge AB. Thus, the light coming from there must be emitted earlier by a time interval $\Delta t = B_o/c$ than the light from FE. If so, the light rays reach the observer (say the camera) simultaneously. But during the time span Δt the the parallelepiped moves on and passes a distance $\Delta x = v \cdot \Delta t = v B_o/c$. Suppose the frontal edges AB and CD are at X_A and X_D, respectively, when the observation (a snapshot) is made (see Fig.2.7b). On account of the length contraction, the edge $X_A X_D$ parallel to the velocity v will appear to the stationary observer to have the length L_o/γ.

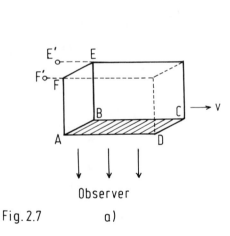

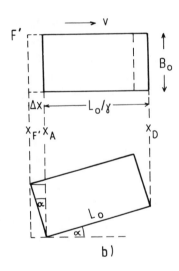

Fig. 2.7   a)   b)

But the image of the forshortened edge $\Delta x = \overline{X_{F'} X_A} = v B_o/c$ so registered by the observer and that of the edge $\overline{X_A X_D} = L_o/\gamma$ (shortened by the length contraction) have precisely the appearence of a parallelepiped at rest which has been rotated by an angle $\alpha = \arcsin(v/c)$. This is so because the projections of the edges $B_o \sin\alpha$ and $L_o \cos\alpha$ give precisely the lengths indicated in the figure if one uses the data $\sin\alpha = v/c$ and $\cos\alpha = \sqrt{1 - v^2/c^2}$. Thus, the parallelepiped will appear rotated by the angle α. Hence the image of the boundary of a sphere which appears rotated in such a manner will look like a circle rather than an ellipse, as was erroneously ascertained in earlier days.

One can also derive the angle of rotation α which was studied in the preceding example with the help of the formula for light aberration. Let Θ be the angle with which light emitted from the object at rest reaches the observer, also at rest. The relation between the angle Θ' pertaining to the case of relative motion and between Θ is given by the relativistic formula for aberration,[1]

$$\sin \theta' = \frac{\sin \theta \sqrt{1-\beta^2}}{1 + \beta \cos \theta}$$

Thus, the moving body appears rotated by an angle $\alpha = \Theta' - \Theta$. If, corresponding to the above example, a snapshot is made in the plane perpendicular to the direction of observation ($\Theta = \pi/2$), then we find $\sin \Theta' = \sqrt{1-\beta^2}$, so that $\cos \alpha = \sqrt{1-\beta^2}$. This agrees with the result which was obtained above in an informal graphic manner.

---

[1]

This aberration formula follows from the transformation equations of the velocity components, cf. equ. (2.19). The sign of β in the denominator arises from the fact that now the observer K' moves with the velocity w = - v relative to the source of light.

Chapter 3

TENSORS

For the covariant formulation of relativistic disciplines (such as mechanics or electrodynamics) one uses certain mathematical entities which possess certain covariance properties under coordinate transformations. One then formulates the physical laws in the form of equations between these entities, called tensors, and thereby one achieves that these equations have the same form in all reference frames which are permitted by the symmetry transformations.

It is useful to define the tensors by their transformational behaviour under general (i.e. also under nonlinear) coordinate transformations. Doing so, the tensor algebra to be derived will be generally valid, i.e. holds true also in non-Euclidean spaces such as those occuring in the theory of gravitation (the general theory of relativity). The tensors to be introduced below in the framework of the pseudo-Euclidean Minkowski space of special relativity theory are to be considered as a special case for a flat space this more general tensor algebra.

Before we begin with this rather abstract introduction of tensors, let us remind ourselves of the graphic definitions of the scalar field, the vector field, and, more generally, of tensor fields.

Scalar fields are functions on space, assigning to each point of space a numerical value. Examples are: temperature, mass, Newtonian potential in mechanics.

Vector fields are functions on space, assigning to each point of space a numerical value and a direction (directed length). Examples are: velocity field, force field, vector potential in electrodynamics.

Tensor fields of rank two are functions on space, assigning linearly to a vector specified at each point of space another vector. Examples are: stress tensor, tensor of inertia, energy-momentum tensor.

In a similar manner, tensor fields of higher rank r assign at each point to a given vector a tensor field of next-lower order r-1, in a linear fashion.

A vector is a tensor of rank one, a scalar is a tensor of rank zero.

3.1 Scalars

We now proceed to define the tensors by means of the behaviour of their components under coordinate transformations. We start with an n-dimensional space, in which one can define coordinate systems in such a way that it is possible to pass from one system $\{x\}$ to another one $\{\bar{x}\}$ with arbitrary continuous, one-to-one coordinate transformations

$$\bar{x}^i = f^i (x^1, x^2, \dots) \quad , \tag{3.1}$$

The inverse transformations are

$$x^k = h^k (\bar{x}^1, \bar{x}^2, \dots) \quad , \quad i, k = 1, 2, \dots n \quad .$$

We allow not only for linear transformations (such as rotations) but also for nonlinear ones. We assume that the first derivatives of these transformations are continuous. Because of this, the functional determinant (the Jacobian) will be a continuous function of space. On account of the invertibility of the transformations, the Jacobian is always nonzero.

Definition: A scalar field (for short: a scalar) is a function $\varphi(P)$ which assigns to each point of the space a number. If one passes from one coordinate system to another one, $x \to \bar{x}$, the value of this function remains unchanged:

$$\bar{\varphi}(P) = \varphi(P) \qquad (3.2)$$

In the space under consideration the point P is characterized by its coordinates. Thus, if the point P has the coordinates x and $\bar{x}$, respectively, in the two coordinate systems, equation (3.2) will read

$$\bar{\varphi}(\bar{x}) = \varphi(x) \qquad (3.3)$$

Thus, the value of the new function $\bar{\varphi}$, which depends on the new coordinates of the point P, must have the same value as has the old function φ, which, in turn, depends on the old coordinates x. If the function $\varphi(x)$ is known, then in this manner the new function $\bar{\varphi}(\bar{x})$ is well defined, since with the inverse transformation of (3.1) x can be expressed in terms of $\bar{x}$:

$$\bar{\varphi}(\bar{x}) := \varphi(h(\bar{x})) \qquad (3.4)$$

It should be observed that, in general, the form of $\bar{\varphi}$ as a function of $\bar{x}$ is not the same as that of φ as a function of x.[1] On the other hand, the numerical value of the field φ at the point P must be the same for all possible coordinates of P which arise from one another by the transformation (3.1). The shorthand notation $\varphi = \bar{\varphi}$ should not lead to misunderstandings, since it simply means that, as in equation (3.2), the functions must be evaluated at the same point.

In this "passive" view of the transformation $x \to \bar{x}$, the point P (or more generally one and the same physical state) is looked at from the two coordinate systems (two observers) $\{x\}$, respectively $\{\bar{x}\}$. The "active" interpretation of symmetry transformations is based on an alternative viewpoint. Here one transforms the physical state (the experimental set-up for the measurement of the state), while the observer's frame of reference stays unchanged. For example, in the case of time reflection, the passive interpretation means the tran-

---

[1] Form invariance of the function means $\bar{\varphi}(\bar{x}) = \varphi(\bar{x})$.

sition to an observer for whom time runs backward. Thus, this transformation is actively realized as the reversal of motion (a transformation of the state).

3.2   Contravariant and Covariant Vector Components

Let us now consider an infinitesimal translation of a point P with coordinates $\{x^i\}$ to a neighbouring point Q with coordinates $\{x^i+dx^i\}$. The translations $\{dx^i\}$ are the components of the infinitesimal vector pointing from P to Q. While this visualization is not necessary, it motivates the following consideration. Let us determine how do the differentials dx^i transform when one goes over from $\{x\}$ to the new coordinate system $\{\bar{x}\}$. To this end, we differentiate the transformation equations (3.1) and obtain right away

$$d\bar{x}^i = \sum_{k=1}^{n} \frac{\partial f^i}{\partial x^k} \, dx^k \tag{3.5}$$

Replacing the function symbols f^i by the barred coordinates $\bar{x}^i$, this can be written in the more convenient form

$$d\bar{x}^i = \sum_{k=1}^{n} \frac{\partial \bar{x}^i}{\partial x^k} \, dx^k \tag{3.6}$$

This transformation law is used to define a vector.

Definition: If an n–tuple of quantities A^i transforms as given by the law (3.6) (as the differentials of the coordinates dx^i) then these A^i represent the contravariant components of a vector:

$$\bar{A}^i(\bar{x}) = \sum_{k=1}^{n} \frac{\partial \bar{x}^i}{\partial x^k} \, A^k(x) \tag{3.7}$$

We speak, for short, of a "contravariant vector". The components of a contravariant vector are labeled by a superscript.

Accordingly, the coordinate differentials dx^i are the contravariant components of the infinitesimal vector from P to Q we talked about earlier.

Clearly, the coordinates x^i of a point in space transform only under linear homogeneous transformations (i.e., rotations)

$$\bar{x}^i = \sum_{k=1}^{n} a_k^i \, x^k \, ,$$

with a_k^i=konst., as the components of a vector. Since the Lorentz transformations are linear, one can, in Minkowski space, introduce the coordinates themselves (not only their differentials) as four-vectors. This is analogous to the definition of the position vector via rotations in the three-dimensional Euclidean space.

A second kind of vector components is defined by considering the behaviour of the partial derivatives of a scalar field ($\partial\varphi/\partial x^i$) under the transformations (3.1). From equation (3.4) it follows with the chain rule that

$$\frac{\partial\bar{\varphi}(\bar{x})}{\partial\bar{x}^i} = \sum_{k=1}^{n} \frac{\partial x^k}{\partial\bar{x}^i} \frac{\partial\varphi(x)}{\partial x^k} \tag{3.8}$$

Here we wrote x^k instead of h^k, which corresponds to our previous agreement (see equ.(3.6)). Using this transformation law, we can define the so-called covariant vector components.

Definition: If an n-tuple of quantities B_k transforms as given by the law (3.8) (as the partial derivatives of a scalar) then these B_k represent the covariant components of a vector:

$$\bar{B}_i(\bar{x}) = \sum_{k=1}^{n} \frac{\partial x^k}{\partial\bar{x}^i} B_k(x) \tag{3.9}$$

We speak, for short, of a covariant vector. In order to distinguish them from the contravariant components, the covariant components are labeled by a subscript.

From the definitions (3.7) and (3.9) the following theorem ensues: The sum of products formed from the contravariant and covariant components of two vectors A and B, i.e. the quantity $\sum_i A^i B_i$ is a scalar.

Indeed, according to the definitions,

$$\bar{S} = \sum_i \bar{A}^i \, \bar{B}_i = \sum_i \sum_l \sum_k \frac{\partial \bar{x}^i}{\partial x^l} \, A^l \, \frac{\partial x^k}{\partial \bar{x}^i} \, B_k \quad .$$

On the other hand,

$$\frac{\partial x^k}{\partial x^l} = \delta^k_l = \sum_i \frac{\partial x^k}{\partial \bar{x}^i} \, \frac{\partial \bar{x}^i}{\partial x^l} \quad , \tag{3.10}$$

where δ^k_l stands for the well known Kronecker symbol. Thus we have

$$\bar{S} = \sum_k \sum_l \delta^k_l \, A^l \, B_k = \sum_k A^k \, B_k = S \quad .$$

In other words, the quantity S stays unchanged under coordinate transformations, i.e., it is a scalar. One calls

$$S = A^k \, B_k \tag{3.11}$$

the scalar product of the vectors A and B. In order to simplify
notations, here and in the sequel the summation symbol is omitted
and we use the summation convention, which means that twice occurring
indices, provided they occur once up and once down, must be summed
over. We also suppressed the indication of the position coordinates,
since it is now clear (see equ.(3.8)) that quantities before and after
the transformation must be computed at the same point, in the sense
of equation (3.2).

As seen from equation (3.10), the transformation of the contravariant vector components A^i goes "contrary" to the transformation
of the covariant components A_i. In view of this, one says that the
transformations (3.7) and (3.9), as well as the corresponding transformation matrices

$$\alpha^i_k = \frac{\partial \bar{x}^i}{\partial x^k} \quad , \quad \beta^i_k = \frac{\partial x^i}{\partial \bar{x}^k} \tag{3.12}$$

are "contragredient" to each other.

3.3   Tensors of Higher Rank

Tensors of higher rank can be obtained by taking products of vectors.

Definition: The contravariant (covariant) components of a tensor with rank r transform as the direct product (tensor product) of r contravariant (covariant) vector components:

$$T^{k_1 k_2 \cdots k_r} \qquad \text{transforms as} \qquad A^{k_1} B^{k_2} \ldots Z^{k_r} \; .$$

Spelling this out in more detail, we have for the contravariant components

$$\overline{T}^{k_1 k_2 \cdots k_r} = \frac{\partial \overline{x}^{k_1}}{\partial x^{m_1}} \frac{\partial \overline{x}^{k_2}}{\partial x^{m_2}} \cdots \frac{\partial \overline{x}^{k_r}}{\partial x^{m_r}} \, T^{m_1 m_2 \cdots m_r} \tag{3.13}$$

Thus, for example, the following three transformation laws are possible for tensors of rank two:

$$\overline{T}^{ik} = \frac{\partial \overline{x}^i}{\partial x^l} \frac{\partial \overline{x}^k}{\partial x^m} \, T^{lm} \qquad\qquad \text{contravariant tensor}$$

$$\overline{T}^i{}_k = \frac{\partial \overline{x}^i}{\partial x^l} \frac{\partial x^m}{\partial \overline{x}^k} \, T^l{}_m \qquad\qquad \text{mixed tensor}$$

$$\overline{T}_{ik} = \frac{\partial x^l}{\partial \overline{x}^i} \frac{\partial x^m}{\partial \overline{x}^k} \, T_{lm} \qquad\qquad \text{covariant tensor}$$

The transformation laws for higher rank tensors of the type (p,q), having p contravariant and q covariant indices as super-, respectively sub-scripts, have similar forms. The rank of such a tensor is $r=p+q$.

3.3.1   Basic Properties of Tensors

Next we discuss some of the important properties of tensors that follow directly from their definition. Two tensors are identical if their components are equal to one another, i.e. if, for all values of the indices, the following relation is fulfilled:

$$A^{ik}_{\ l} = B^{ik}_{\ l}$$

It suffices that this holds in a particular coordinate system: due to the definition of tensors, the equality follows to be valid also in all coordinate systems. Consequently, if one formulates physical laws as equations beween tensors, these laws will hold in the same form in all coordinate system which arise from each other by means of the basic transformations (3.1).[1]

Since the definition of tensors is linear and homogeneous in the tensor components, the following algebraic rules hold true:

The sum of like tensors is again a tensor of the same type,

$$A^{ik}_{\ l} + B^{ik}_{\ l} = C^{ik}_{\ l}$$

The product of a tensor with a scalar (i.e., the multiplication of each component with a scalar) yield again a tensor.

The direct product of two tensors gives a tensor of correspondingly higher rank (in general of mixed rank),

$$A^{ik}_{\ l} B^{mn} = G^{ikmn}_{\ l} \ .$$

A special case is the direct product of, say, three vectors,

$$T^{ikl} = A^i B^k C^l \ .$$

More generally, every tensor may be written as a sum of such direct products of vectors.[2]

A tensor is said to be symmetric in a certain pair of indices if

$$T_{ik} = T_{ki}$$

---

[1]

In quantum theory, apart of tensors one also encounters spinors. These correspond to other possible representations of the Lorentz group. While tensor fields correspond to quanta with integral spin values 0, 1, 2,... spinor fields describe quanta with half-integral spin 1/2, 3/2... .

[2] See, for example, R. Adler, M. Bazin, and M. Schiffer: Introduction to General Relativity, 2nd ed. (Mc Graw-Hill, New York, 1975),p.25.

and skew-symmetric (or antisymmetric) if

$$T_{ik} = -T_{ki}$$

In particular, every tensor of rank 2 can be split into the sum of a symmetric and an antisymmetric tensor:

$$T_{ik} = \frac{1}{2} (T_{ik} + T_{ki}) + \frac{1}{2} (T_{ik} - T_{ki})$$

For the case of a three-dimensional space (n=3), a symmetric tensor has 6, an antisymmetric tensor 3 essential components. Because of this, the customary vector product in a three-dimensional space (which, strictly speaking, represents an antisymmetric tensor of rank 2) may be written down again as a vector (with 3 components).

3.3.2 Contractions

Tensors can be contracted, i.e., one can reduce their rank. For example, if one forms from the tensors A^{ik} and B_1 the quantity

$$C^i = A^{ik} B_k \quad ,$$

where, according to our agreement, summation over the sub- and superscript k is understood, then the C^i are contravariant components of a vector. To prove this, one uses the transformation properties of A^{ik} and B_k. With the notations of equ.(3.12) one obtains

$$\bar{C}^i = \bar{A}^{ik} \bar{B}_k$$
$$= \alpha^i_l \alpha^k_m A^{lm} \beta^n_k B_n$$

Because of (3.10) it follows that

$$\bar{C}^i = \alpha^i_l \delta^n_m A^{lm} B_n$$
$$= \alpha^i_l A^{lm} B_m = \alpha^i_l C^l \quad .$$

But this is precisely the transformation law for vectors.

The formation of the scalar product of two vectors (see equ. (3.11)) is yet another example of contraction. Contraction of a superscript with a subscript of a mixed tensor of the type (p,q) results in a tensor of type (p-1, q-1). Thus, for example, contraction of the tensor T_1^k leads to a scalar,

$$T_k^k = S \ .$$

3.3.3 The Quotient Theorem

This theorem is useful for determining the tensor character of a given expression. It will suffice to illustrate the theorem by a simple example; its generalization will be obvious.

Let A_{ik} be a quantity with n^2 components and suppose that $A_{ik} X^k$, where X^k is an arbitrary contravariant vector, represents a covariant vector. Then A_i is a covariant tensor of rank 2.

According to the assumption, $A_{ik} X^k$ must transform as a covariant vector, so that (with the notations of (3.12)) we have

$$\bar{A}_{ik} \bar{X}^k = \beta_i^l A_{lm} X^m$$

For the contravariant vector components $\bar{X}^k$ on the left-hand side we take into account the transformation (3.7),

$$\bar{A}_{ik} \alpha_m^k X^m = \beta_i^l A_{lm} X^m$$

Since this must hold for arbitrary vectors X^m, comparison of coefficients yields

$$\bar{A}_{ik} \alpha_m^k = \beta_i^l A_{lm}$$

Multiplying both sides by β_n^m and summing over m, we finally get, with (3.10),

$$\bar{A}_{in} = \beta_i^l \beta_n^m A_{lm}$$

This shows that the quantities A_{lm} transform as the covariant components of a rank 2 tensor.

With an analogous kind of proof one verifies the following: If B is an arbitrary tensor and if the product AB=C results in tensor C, then A is a tensor. Here the product may involve an arbitrary number of contractions of upper and lower indices between A and B, but not within A or B.

A further example of applying the theorem will demonstrate that δ^i_k is a tensor. To this end, we write the scalar product of two arbitrary vectors in the form

$$S = X^i Y_i = (\delta^i_k X^k) Y_i$$

Since Y_i is an arbitrary vector, the quotient theorem tells us that $\delta^i_k X^k$ is a vector. Applying now once again the quotient theorem, the tensor property of δ^i_k follows immediately. Thus, under the general transformations (3.1), the Kronecker symbol transforms as a mixed tensor.

3.3.4 Relative Tensors

The absolute tensors we considered so far, form a special class of the somewhat more generally defined relative tensors. These are quantities that transform like tensors except that in their transformation law there also appears some power of the functional determinant of the transformation.

Definition: A set of quantities obeying the transformation law

$$\bar{R}^{k_1 \cdots k_r}_{m_1 \cdots m_s} = \frac{\partial \bar{x}^{k_1}}{\partial x^{l_1}} \cdots \frac{\partial \bar{x}^{k_r}}{\partial x^{l_r}} \frac{\partial x^{n_1}}{\partial \bar{x}^{m_1}} \cdots \frac{\partial x^{n_s}}{\partial \bar{x}^{m_s}} R^{l_1 \cdots l_r}_{n_1 \cdots n_s} \left| \frac{\partial x}{\partial \bar{x}} \right|^w ,$$

where $\left| \partial x / \partial \bar{x} \right|$ stands for the functional determinant of the transformation, is called a relative tensor with weight w.

The algebraic operations for relative tensors are analogous to those for ordinary tensors. But one must take note of the fact that, for example if one forms the direct product of relative tensors, the weight will add.

For $w=0$ we get back the absolute tensors discussed above. A relative tensor with weight $+1$ is called a tensor density. This terminology is based on the fact that under a general coordinate transformation $x \rightarrow \bar{x}$ of the volume element the functional determinant will appear as a factor:

$$d^n \bar{x} = d^n x \left| \frac{\partial \bar{x}}{\partial x} \right| = d^n x \left| \frac{\partial x}{\partial \bar{x}} \right|^{-1} \tag{3.14}$$

Thus, the volume element $d^n x$ is a relative scalar with weight -1. Hence, the product of the volume element with a tensor density transforms as an ordinary tensor. In particular, the integration over a scalar density $A(x)$ yields an (absolute) scalar S,

$$S = \int d^n x \, A(x)$$

Another interesting example of tensor densities is the Levi–Civita symbol $\varepsilon^{ikl\cdots}$. This entity, antisymmetrical in all index pairs, is defined in the following way:

$$\varepsilon^{i_1 i_2 \cdots i_n} = \begin{cases} + 1 \text{ for indices which are an even permutation of } 12\ldots n \\ - 1 \text{ for } \quad " \quad " \quad " \quad " \text{ odd} \quad " \quad " \\ 0 \text{ for identical indices} \end{cases}$$

Consequently, the quantity

$$\frac{\partial \bar{x}^{i_1}}{\partial x^{k_1}} \frac{\partial \bar{x}^{i_2}}{\partial x^{k_2}} \cdots \frac{\partial \bar{x}^{i_n}}{\partial x^{k_n}} \, \varepsilon^{k_1 k_2 \cdots k_n}$$

is antisymmetric in all index pairs, hence proportional to $\varepsilon^{i_1 \cdots i_n}$. To find the factor of proportionality, take for the indices $i_1 i_2 \ldots i_n$ the numbers $1 \, 2 \ldots n$. One then finds that the factor equals the determinant

$$\frac{\partial \bar{x}^1}{\partial x^{k_1}} \frac{\partial \bar{x}^2}{\partial x^{k_2}} \cdots \frac{\partial \bar{x}^n}{\partial x^{k_n}} \, \varepsilon^{k_1 k_2 \cdots k_n} = \left| \frac{\partial \bar{x}}{\partial x} \right|$$

so that

$$\frac{\partial \bar{x}^{i_1}}{\partial x^{k_1}} \frac{\partial \bar{x}^{i_2}}{\partial x^{k_2}} \cdots \frac{\partial \bar{x}^{i_n}}{\partial x^{k_n}} \, \varepsilon^{k_1 k_2 \cdots k_n} = \left| \frac{\partial \bar{x}}{\partial x} \right| \, \varepsilon^{i_1 i_2 \cdots i_n}$$

Hence, $\varepsilon^{i_1 i_2 \cdots i_n}$ is a tensor density.

One can use the Levi–Civita tensor to construct antisymmetric tensors. If, for example, T_{ik} is a tensor, then

$$T_{ik} \, \varepsilon^{klm\cdots} = R_i^{lm\cdots}$$

represents a relative tensor with weight $+1$, which is antisymmetric in its contravariant indices. In particular, for a three-dimensional space the construction

$$A_i \varepsilon^{ikl} = T^{kl}$$

gives a skew-symmetric tensor of rank 2, whenever A_i is a relative vector with weight -1. In this manner, a unique correspondence between A_i and T^{kl} is established. The customary vector product $\vec{a} \times \vec{b}$ may be looked upon as either one of these quantities.

3.4   The Metric Tensor

For the three-dimensional Euclidean space the squared distance between two infinitesimally near points (i.e., the line element ds) is given, in Cartesian coordinates, by the expression

$$ds^2 = dx^2 + dy^2 + dz^2 \tag{3.15}$$

This is a scalar. Introducing curvilinear coordinates, $\vec{r} = \vec{r}(u^1, u^2, u^3)$ then, in consequence of

$$d\vec{r} = \frac{d\vec{r}}{du^i} du^i \quad , \, i = 1, 2, 3$$

one gets

$$ds^2 = \frac{d\vec{r}}{du^i} \cdot \frac{d\vec{r}}{du^k} du^i \, du^k = g_{ik} \, du^i \, du^k \tag{3.16}$$

Clearly, the coefficients g_{ik} determine the metric. Later we shall see that these g_{ik} form components of a tensor. Therefore g_{ik} is called the metric tensor. For Cartesian coordinates (cf.(3.15)) $g_{ii} = 1$ and $g_{ik} = 0$ for $i \neq k$. A less trivial example is given by the spherical coordinates $\{r, \Theta, \Phi\}$. Here the squared infinitesimal distance is given by

$$ds^2 = dr^2 + r^2 d\Theta^2 + r^2 \sin^2\Theta \, d\Phi^2$$

Comparing with (3.16) one then obtains the corresponding metric tensor

$$g_{ik} = \begin{pmatrix} 1 & 0 & 0 \\ 0 & r^2 & 0 \\ 0 & 0 & r^2 \sin^2 \theta \end{pmatrix}$$

In general, the connection between the scalar ds^2 and the coordinate differentials dx^i in an n-dimensional Riemann space is given by the quadratic form

$$ds^2 = g_{ik}(x)\, dx^i\, dx^k, \qquad i,k = 1,\ldots,n \qquad (3.17)$$

where the metric tensor g_{ik} may depend on the coordinates.

First we convince ourselves that under the transformation (3.1) g_{ik} behaves as a tensor. Let dx be an arbitrary vector, composed of the two arbitrary vectors $dx_{(1)}$ and $dx_{(2)}$,

$$dx = dx_{(1)} + dx_{(2)}$$

Correspondingly, the scalar ds^2 may be written as a sum of terms:

$$ds^2 = g_{ik}\, dx^i\, dx^k$$

$$= g_{ik}\, dx^i_{(1)} dx^k_{(1)} + g_{ik}\, dx^i_{(2)} dx^k_{(2)} + 2 g_{ik}\, dx^i_{(1)} dx^k_{(2)}$$

According to the definition (3.17), ds^2, $ds^2_{(1)}$, and $ds^2_{(2)}$ are scalars, so that the last term on the right-hand side must be also a scalar. Since $dx^i_{(1)}$ and $dx^k_{(2)}$ are two different arbitrary vectors, the quotient theorem tells us that g_{ik} is a tensor of rank two, i.e., it has the transformation law

$$\bar{g}_{ik} = \frac{\partial x^l}{\partial \bar{x}^i}\, \frac{\partial x^m}{\partial \bar{x}^k}\, g_{lm} \qquad (3.18)$$

One should note that the dx^i, dx^k in equation (3.17) are only different components of one and the same vector. Therefore the quotient theorem cannot be directly applied. Only the decomposition used above allows for the fulfilment of applicability.

Since an antisymmetric contribution to g_{ik} would not alter the value of the squared line element ds^2 in equ.(3.17), we can assume that g_{ik} is symmetric.

The contraction of tensors, as discussed above, permits the following interpretation of equation (3.17). One clearly has here contracted the tensor g_{ik} with the contravariant vector components dx^i to yield the covariant components dx_k, and subsequently dx_k is contracted with dx^k to give the scalar $ds^2=dx_k dx^k$. In other words, contracting with g_{ik} we can change the contravariant components of a tensor into its covariant components, without changing the tensor as such. To put it succinctly, contraction with g_{ik} pulls down (lowers) upper indices.

It is now justified to ask: is there a contravariant tensor which makes it possible a corresponding pulling up (raising) of indices? To this end, one introduces the contravariant tensor g^{ik}, which is the inverse of g_{ik}:

$$g^{il} g_{lk} = \delta^i_k \qquad (3.19)$$

This determines the components of g^{ik}. The inverse matrix of g_{ik} is

$$g^{ik} = \frac{\Delta^{ki}}{g} \qquad (3.20)$$

Here Δ^{ki} stands for the algebraic complement of g_{ki} and g is the determinant of the matrix g_{ik}. For example, in the case of spherical coordinates one finds that

$$g^{ik} = \begin{pmatrix} 1 & 0 & 0 \\ 0 & 1/r^2 & 0 \\ 0 & 0 & 1/r^2\sin^2\theta \end{pmatrix}$$

The tensor nature of g^{ik} follows from its definition from g_{ik} and δ^i_k, as represented in equation (3.19). To convince oneself that contraction with g^{ik} effects a raising of indices, one first constructs the quantity

$$A^i = g^{il} A_l .$$

If these are the contravariant components of a vector, then the contraction with g_{ik} should yield the covariant components. As expected, (3.19) leads to the desired result,

$$g_{ki} A^i = g_{ki} g^{il} A_l = A_k$$

Contraction with g_{ik} or g^{ik} may be applied to arbitrary tensor components. One only has to watch out for the position of the indices. The position of free indices must be preserved. Examples are:

$$T^i_{\ k} = g_{kl} T^{il} \qquad \text{or} \qquad T^{ik} = g^{il} T_l^{\ k} = g^{il} g^{km} T_{lm} \ .$$

The tensor components so associated with each other by g_{ik} or g^{ik} refer to different bases which are constructed from contravariant and covariant basis vectors respectively. It is conventional to give all these (inequivalent) tensors the same name (e.g.T), distinguishing them only by the positions of their indices. Whichever components one uses, the relation between geometrical or physical quantities, as expressed by a tensor equation, always remains the same.

A modern approach to tensor analysis stresses the geometrical nature of tensors rather than the transformation properties of their components. Accordingly, vectors (contravariant vectors) and covectors (covariant vectors) are, in general, different coordinate-independent geometrical objects. Therefore, in modern terminology the contravariant vector is called briefly a vector, whereas, in contrast, the covector is termed a one-form.[1]

The metric tensor may also be used to convert a tensor into a relative tensor (or vice versa), and more generally to change the weight of a relative tensor. Indeed, taking the determinants of the matrices in equation (3.18), one finds, with $g=\det(g_{ik})$ that

$$\sqrt{\bar{g}} = \left| \frac{\partial x}{\partial \bar{x}} \right| \sqrt{g} \ .$$

In other words, $\sqrt{g}$ is a relative tensor with weight +1 (a scalar density). Thus, the factor $\sqrt{g}$ may be used to raise the weight of a relative tensor by 1.

---

[1] For more details see B.F. Schutz: Geometrical Methods in Mathematical Physics (Cambridge University Press, Cambridge, 1980).

3.5   Differentiation of Tensor Fields

As we saw earlier, the partial derivatives of a scalar field represent a covariant vector. However, in contrast, the derivatives of a vector field or of a tensor field of higher rank do not lead in general to a new tensor. For example, differentiating equation (3.7) one gets

$$\frac{\partial \bar{A}^i}{\partial \bar{x}^k} = \frac{\partial^2 \bar{x}^i}{\partial x^m \partial x^n} \frac{\partial x^n}{\partial \bar{x}^k} A^m + \frac{\partial \bar{x}^i}{\partial x^m} \frac{\partial x^n}{\partial \bar{x}^k} \frac{\partial A^m}{\partial x^n} \tag{3.21}$$

The deviation from tensorial behaviour is exhibited by the first term. This is nonzero when the transformation matrix $\partial \bar{x}^i / \partial x^k$ itself depends on the coordinates, i.e., if the transformation is nonlinear. For linear transformations, there is no problem. However, one can use the metric tensor g_{ik} to define a generalized (covariant) differentiation in Riemannian space with arbitrary transformations, such that when it is applied to a tensor, one obtains again a tensor. This is achieved with the so-called Christoffel symbols (of second kind),[1] given by

$$\Gamma^i_{kl} = \frac{1}{2} g^{im} \left(\frac{\partial g_{mk}}{\partial x^l} + \frac{\partial g_{ml}}{\partial x^k} - \frac{\partial g_{kl}}{\partial x^m} \right)$$

Because the appearance of the derivatives of g_{ik} this quantity does

---

[1] More generally, the quantities Γ^i_{kl} are introduced as the so-called "coefficients of affine connection" in order to define a rule for parallel transport of vector fields on a differentiable manifold wherein no metric g_{ik} is yet defined. If, after introducing a metric, one requires that the scalar product of two vectors be invariant under parallel displacement, then it is possible to express the Γ^i_{kl} in terms of the metric tensor and its first derivatives, as hinted at above. We refer the reader to the topical textbooks. See, for example, D. Laugwitz: Differential and Riemannian Geometry (Academic Press, New York, 1965); D. Lovelock, H. Rund: Tensors, Differential Forms, and Variational Principles (J. Wiley, New York, 1975).

not transform like a tensor. Rather, its transformation law is:

$$\bar{\Gamma}^i_{kl} = \frac{\partial \bar{x}^i}{\partial x^m} \frac{\partial x^n}{\partial \bar{x}^k} \frac{\partial x^p}{\partial \bar{x}^l} \Gamma^m_{np} - \frac{\partial x^n}{\partial \bar{x}^k} \frac{\partial x^p}{\partial \bar{x}^l} \frac{\partial^2 \bar{x}^i}{\partial x^n \partial x^p}$$

Using this and (3.7), one calculates that

$$\bar{\Gamma}^i_{kl} \bar{A}^l = \frac{\partial \bar{x}^i}{\partial x^m} \frac{\partial x^n}{\partial \bar{x}^k} \Gamma^m_{np} A^p - \frac{\partial^2 \bar{x}^i}{\partial x^n \partial x^p} \frac{\partial x^n}{\partial \bar{x}^k} A^p \tag{3.22}$$

The second term equals, up to the sign, the perturbing term in equation (3.21). Adding equations (3.21) and (3.22), these terms cancel and one obtains

$$\frac{\partial \bar{A}^i}{\partial \bar{x}^k} + \bar{\Gamma}^i_{kl} \bar{A}^l = \frac{\partial \bar{x}^i}{\partial x^m} \frac{\partial x^n}{\partial \bar{x}^k} \left(\frac{\partial A^m}{\partial x^n} + \Gamma^m_{np} A^p \right)$$

In this manner one arrives at the definition of the covariant derivative:

$$A^i_{;k} := \frac{\partial A^i}{\partial x^k} + \Gamma^i_{kl} A^l \tag{3.23}$$

If A^i is a contravariant vector, then its covariant derivative $A^i_{;k}$ transforms as a mixed rank 2 tensor:

$$\bar{A}^i_{;k} := \frac{\partial \bar{x}^i}{\partial x^m} \frac{\partial x^n}{\partial \bar{x}^k} A^m_{;n}$$

Since the partial derivatives of a scalar already yield a vector, one expects that the covariant derivative of a scalar should coincide with the simple partial derivative. If one therefore applies covariant derivation to the scalar product of two arbitrary vectors and applies the chain rule, one obtains

$$(A^i B_i)_{;k} = A^i_{;k} B_i + A^i B_{i;k} = \frac{\partial}{\partial x^k} (A^i B_i)$$

With definition (3.23) it now follows from this formula the expression of the generalized derivation of the covariant components B_i of a vector:

$$B_{i\,;k}: = \frac{\partial B_i}{\partial x^k} - \Gamma^n_{ik} B_n \tag{3.24}$$

On account of the negative sign in (3.24) those terms in the covariant derivation of a scalar product which contain the Christoffel symbols, exactly cancel.

Based on the definition of higher tensors as direct products of vectors, it is clear how these rules for covariant derivation can be extended to tensors of higher rank.

With the aid of covariant derivation one can formulate tensor analysis in curved spaces or in arbitrary coordinate systems. This is very important in the framework of general relativity (theory of gravitation). For linear transformations, such as the Lorentz transformations in Minkowski space, the components of the metric tensor do not depend on the coordinates, so that $\Gamma^i_{kl}=0$; i.e., the covariant derivative becomes the simple partial derivative. Consequently, in the special theory of relativity, assuming that one uses Cartesian coordinates, the partial derivatives of a tensor yield again a tensor with respect to Lorentz transformations. – However, if one switches in the Minkowski space to curvilinear coordinates, by using some non-linear transformation (for example, if one introduces spatial polar coordinates), then the first term in (3.21) is not zero, and the partial derivatives do not represent a tensor with respect to this transformation. Only the covariant derivatives give tensors in this case.

These remarks should suffice to indicate the general framework into which the simple tensor analysis used in the special theory of relativity fits.

3.6   Vectors in Euclidean Space

So as to visualize the contravariant and covariant components of a vector or tensor, which we introduced in abstracto above, we now consider the familiar case of the three-dimensional Euclidean space. We shall see forthwith that the covariant and contravariant components of vectors, associated to each other by the metric tensor g_{ik}, precisely correspond to the two varieties which define the components of a vector in Euclidean space.

Let us begin with a non-rectilinear coordinate system which we define by the unit vectors $\vec{e}_1$, $\vec{e}_2$, $\vec{e}_3$ that carry distinguishing subscripts.[1] The decomposition of a vector $\vec{a}$ into three vectors, parallel to these three directions, reads

$$\vec{a} = a^1 \vec{e}_1 + a^2 \vec{e}_2 + a^3 \vec{e}_3 = a^i \vec{e}_i$$

If, however, one projects the vector $\vec{a}$ perpendicularly to the directions $\vec{e}_i$, the usual scalar product leads to the projections

$$a_i = \vec{a} \cdot \vec{e}_i$$

Specification of the a_i determines the vector $\vec{a}$. For the systematic notation of the projections we always use the subscripts which originate from the convention regarding the indices of the unit vectors. We may call these projections the covariant components of the vector $\vec{a}$. If so, then the quantities a^i which occur when the vector was decomposed parallel to the unit vectors, represent the contravariant components of $\vec{a}$. To see this, we express the scalar product of two vectors $\vec{a}$ and $\vec{b}$ by these components:

$$\vec{a} \cdot \vec{b} = (a^i \vec{e}_i) \cdot (b^k \vec{e}_k) = \vec{e}_i \cdot \vec{e}_k\, a^i b^k$$

---

[1] We could also start with a basis that uses superscripts. But one has to take care of the position of indices in the sequel, once a decision is made.

With the definition

$$g_{ik} := \vec{e}_i \cdot \vec{e}_k$$

this can be rewritten in the already known form

$$\vec{a} \cdot \vec{b} = g_{ik} \, a^i \, b^k$$

Since the factors $\vec{e}_i$ and $\vec{e}_k$ commute, g_{ik} is symmetric.

We can perform the decomposition of the vectors in the scalar product also differently, viz.

$$\vec{a} \cdot \vec{b} = \vec{a} \cdot (b^k \vec{e}_k) = \vec{a} \cdot \vec{e}_k \, b^k = a_k \, b^k$$

Thus, comparison of these two representations gives

$$a_k = g_{ik} \, a^i$$

since the vector $\vec{b}$ is arbitrary. This shows that g_{ik} may be identified with the metric tensor that converts the contravariant components a^i of the vector $\vec{a}$ into its covariant components. Since $\vec{a}$ and $\vec{b}$ are arbitrary, the tensor property of g_{ik} is evident.

Using the abbreviations (3.12), the transformation law of the contravariant components becomes

$$\bar{a}^i = \alpha_k^i \, a^k$$

The transformation law of the basis vectors $\vec{e}_i$ may be found by using the covariant components,

$$\bar{a}_i = \beta_i^n \, \vec{a} \cdot \vec{e}_n = \vec{a} \cdot \beta_i^n \, \vec{e}_n$$

which means that

$$\bar{\vec{e}}_i = \beta_i^n \, \vec{e}_n$$

Using now the identity $\alpha_k^i \beta_i^n = \delta_k^n$ (see equ.(3.10)), it now follows that

$$\vec{e}_k = \alpha_k^i \, \overline{\vec{e}}_i$$

Thus, in this sense, the transformation of the contravariant components a^i goes in "the opposite way" ("contra") to the transformation of the basis $\{\vec{e}_i\}$. On the other hand, the covariant components a_i transform in "the same way" ("conformly") as does the basis $\{\vec{e}_i\}$. All this explains the choice of the terminology "contra" and "co"-variant.

According to equation (3.19), the g_{ik} also define the contravariant components g^{ik} of the metric tensor. With it, one can raise the indices of the basis vectors and one is led to the contravariant basis

$$\vec{e}^i = g^{ik} \, \vec{e}_k$$

In agreement with the previous, the inner product now reads

$$\vec{a} \cdot \vec{b} = (a_i \, \vec{e}^i) \cdot (b_k \, \vec{e}^k) = \vec{e}^i \cdot \vec{e}^k \, a_i \, b_k = g^{ik} \, a_i \, b_k$$

What is the relation between the basis vectors $\vec{e}^i$ and $\vec{e}_k$? Forming the inner product we get

$$\vec{e}^i \cdot \vec{e}_k = g^{in} \vec{e}_n \cdot \vec{e}_k = g^{in} g_{nk} = \delta_k^i$$

These relations are satisfied by the expressions

$$\vec{e}^1 = \frac{\vec{e}_2 \times \vec{e}_3}{\vec{e}_1 \cdot \vec{e}_2 \times \vec{e}_3} \quad , \quad \vec{e}^2 = \frac{\vec{e}_3 \times \vec{e}_1}{\vec{e}_1 \cdot \vec{e}_2 \times \vec{e}_3} \quad , \quad \vec{e}^3 = \frac{\vec{e}_1 \times \vec{e}_2}{\vec{e}_1 \cdot \vec{e}_2 \times \vec{e}_3}$$

The basis vectors of one vector system are orthogonal to the coordinate planes of the other vector system. The basis vectors $\{\vec{e}^i\}$ span the space dual to the space of the $\{\vec{e}_i\}$. In the lattice theory of solids these spaces are termed "reciprocal". One talks of reciprocal lattices.

The metric tensor determines the magnitude of the basis vectors and the angles between them (which need not be a right angle). Thus, from the definitions

$$g^{ik} = \vec{e}^i \cdot \vec{e}^k \quad , \quad g_{ik} = \vec{e}_i \cdot \vec{e}_k$$

one obtains for the magnitude of the contravariant (covariant) basis vectors:

$$|\vec{e}^i| = \sqrt{g^{ii}} \quad , \quad |\vec{e}_i| = \sqrt{g_{ii}}$$

The angles between the basis vectors are then

$$\cos(\vec{e}_i, \vec{e}_k) = \frac{\vec{e}_i \cdot \vec{e}_k}{|\vec{e}_i||\vec{e}_k|} = \frac{g_{ik}}{\sqrt{g_{ii}\, g_{kk}}} \quad , \quad \cos(\vec{e}^i, \vec{e}^k) = \frac{g^{ik}}{\sqrt{g^{ii}\, g^{kk}}}$$

In Fig.3.1 we can visualize the above facts and statements. The figure shows the components of a vector $\vec{a}$ which lies in the plane of the basis vectors $\vec{e}_1$, $\vec{e}_2$ and also of $\vec{e}^1$, $\vec{e}^2$.

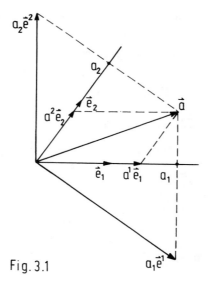

Fig. 3.1

The covariant basis vectors $\vec{e}_1$ and $\vec{e}_2$ are assumed to have unit length. Then the covariant components a_i are obtained as the orthogonal projections of $\vec{a}$ onto the directions $\vec{e}_i$. The relations $\vec{e}^1 \cdot \vec{e}_1 = 1$ etc. which follow from (3.25) do not imply that the $\vec{e}^i$ are parallel to the $\vec{e}_i$, because the basis vectors $\vec{e}^i$ need not have unit length. The angle between the basis vectors having the same indices is determined by the formula (no summation over i)

$$\cos(\vec{e}^i, \vec{e}_i) = \frac{1}{|\vec{e}^i||\vec{e}_i|}$$

If, for example, the basis vectors $\vec{e}_i$ are unit vectors but not mutually orthogonal, then it follows from (3.25) that the dual basis vectors $\vec{e}^i$ do not have unit length. But then $\cos(\vec{e}^i, \vec{e}_i) \neq 1$, so that the corresponding basis vectors are not parallel.

For a Cartesian coordinate system with unit vectors for its base, one clearly has $g_{ik}=1$ for i=k and $g_{ik}=0$ for i≠k, so that $a^i=a_i$, which is to say, contravariant and covariant components coincide. Thus, in this case one need not distinguish between kinds of components.

As we saw, in an Euclidean space the coefficients g_{ik} are interpreted geometrically as the inner product of basis vectors. In differential geometry (for example in the study of a surface in space), one has

$$ds^2 = g_{ik}(x)\,dx^i\,dx^k$$

where the dx^i stand for the infinitesimal changes of the coordinates on the surface. Here one can locally interpret the $g_{ik}(x)$ as the scalar product of the tangent vectors of the coordinate lines (see equ. (3.16)). Using these vectors as a basis, one then has locally defined a tangential Euclidean space.

If one component of the metric tensor in an Euclidean space is negative, one talks of a pseudo-Euclidean space. This applies to the Minkowski space of special relativity.

Chapter 4

FORMULATION OF RELATIVITY THEORY
IN MINKOWSKI SPACE

4.1 The Four-Dimensional Minkowski Space

As already mentioned on p.34, one arrives at Minkowski space, and thereby at a rational formulation of special relativity theory, if one enlarges the set of spatial coordinates to a four-dimensional space-time continuum,

$$x^\mu = (x^0, \vec{x}) , \quad x^1 = x, \quad x^2 = y, \quad x^3 = z$$

which is achieved by adding the coordinate $x^0 = ct$. Now the requirement of invariance under the Galilei group (including rotations) in the three-dimensional space must be replaced by the requirement of invariance under the Lorentz transformations, which can be looked upon as orthogonal transformations in Minkowski space, analogous to rotations in three dimensions. Physical laws are then formulated as covariant equations between four-dimensional tensor entities.

Using the metric tensor $g_{\mu\nu}$ and the above definition of x^μ, the separation between world points (cf. equ.(2.12) in this space may be written as

$$s_{12}^2 = g_{\mu\nu} (x_{(2)}^\mu - x_{(1)}^\mu)(x_{(2)}^\nu - x_{(1)}^\nu) , \quad \mu, \nu = 0, 1, 2, 3 .$$

This quantity is invariant under Lorentz transformations. As seen by comparing with (2.12), the components of the metric tensor in Minkowski space are

$$g_{\mu\nu} = \begin{pmatrix} 1 & 0 & 0 & 0 \\ 0 & -1 & 0 & 0 \\ 0 & 0 & -1 & 0 \\ 0 & 0 & 0 & -1 \end{pmatrix} \tag{4.1}$$

Therefore, the square of the "length" of a four-vector x^{μ} is

$$s^2 = x^2 = g_{\mu\nu} x^{\mu} x^{\nu} = x^{o^2} - \vec{x}^2 \tag{4.2}$$

The covariant components of a four-vector x are obtained by contraction with the metric tensor (4.1),

$$x_{\mu} = g_{\mu\nu} x^{\nu}$$

While the contravariant and covariant temporal components of a four-vector are the same, the corresponding spatial components differ in sign:

$$x_0 = x^0 , \quad x_n = -x^n , \quad n = 1,2,3 .$$

Because of the negative signs in the metric tensor, the inner product $x^{\mu} x_{\mu}$ (and hence the squared separation of two events) is no longer positive definit (as it is in the three-dimensional Euclidean space); it can be positive, negative, or zero. In this manner, separations (and more generally vectors) in Minkowski space can be classified into three categories. In particular, for the propagation of light signals we have, in all coordinate systems,

$$x_{\mu} x^{\mu} = 0$$

(see equ.(2.9). We call any four-vector with this property a null vector. If one marks x^o and x^1 in the same units on the coordinate axes in the $x^o x^1$ plane, then the world line corresponding to light signals is a straight line x=ct (x=-ct) with slope 45° (135°), passing

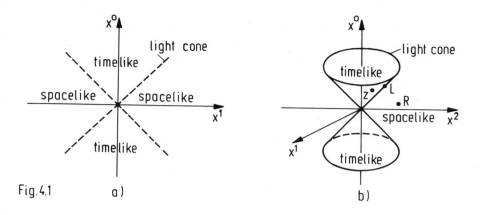

Fig.4.1   a)   b)

through the origin (see Fig.4.1a).[1] Adding x^2 as a second spatial coordinate, then rotating around the x^o axis makes the straight line to become the mantle of a cone (Fig.4.1b). For short, we call this object the "light cone". Since for particles with nonzero mass we always have $v < c$, the world lines of such particles that go through the origin must stay inside the light cone.

a) Let us first consider an event which lies outside the light cone, i.e., let the coordinates x^μ refer to point R in Fig.4.1b. For such points

$$s^2 = x^{o^2} - \vec{x}^2 < 0$$

Since s^2 is invariant, this holds in all inertial systems. In particular, it is possible to chose a frame of reference K' such that $(x'^o)^2 = 0$. But it is not possible to find one where $\vec{x}'^2 = 0$. This would contradict $s^2 < 0$. Therefore, in all inertial systems the world points O and R have a spacelike separation. The world point R is "absolutely distant" from O. Separations (intervals) with $s^2 < 0$ are called "spacelike".

_____

[1] The orthogonality of the coordinate axes in the $x^o x$ plane, as chosen for convenience here, is merely a matter of convention. Passing to another coordinate system K' with a Lorentz transformation, the new axes in the $x^o{'}x'$ plane are no longer orthogonal (cf. Chapter 4.4).

One uses this terminology for all four-vectors such that $A^\mu A_\mu < 0$. However, the concepts of simultaneity, earlier and later, are relative concepts when we talk about events that lie outside the light cone (such as the world point R). This is so because by passing over to some suitable other frame of reference one can arrange that events such as R occur earlier or later than the event point O. As already noted above, there also exists a system in which R is simultaneous with O ($x'^o=0$).

b) Next we consider events Z which lie inside the light cone (see Fig.4.1b). For these world points we have

$$s^2 = x^{o2} - \vec{x}^2 > 0$$

In this case there is no inertial system in which we would have $x'^o=0$, but it is possible to achieve that $\vec{x}'^2=0$. Thus in all inertial systems the events O and Z have a temporal (timelike) separation. Separations with $s^2 > 0$ are called "timelike". More generally, four-vectors A^μ such that $A^\mu A_\mu > 0$ are called timelike vectors. Two events separated by a timelike interval, cannot occur simultaneously in any frame of reference. For $x^o > 0$ the events of type Z occur always later than the event O. The condition $x^o > 0$ cannot be changed by a Lorentz transformation to $x^o < 0$. This is so because, if it could be done, then, on account of continuity, one could also have $x^o=0$ which is impossible if $s^2 > 0$. For this reason, the region $s^2 > 0, x^o > 0$ (inside of the positive (forward) light cone)is called "absolute future". Similarly,the region $s^2 > 0, x^o < 0$ (the inside of the negative (backward) light cone) is called the "absolute past" relative to O, since events in this region occur in all reference systems before the event O. This affair of states is visualized in Fig.4.2.

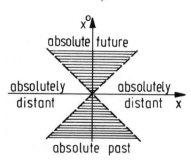

Fig. 4.2

c) Finally, we may encounter the already mentioned case, where an event L lies on the light cone. Then

$$s^2 = x^{o2} - \vec{x}^2 = 0$$

holds in all inertial systems. These events, or more generally, four-vectors with "length" zero, are called "lightlike". World lines of massless particles (which propagate with the speed of light) lie on the light cone.

We see that world points on the light cone correspond to events that can be connected to O by a light signal. Since effects cannot propagate with a velocity exceeding that of light, any effect originating at O can reach only those world points which lie inside the positive light cone or on the cone surface. Put otherwise: a causal connection between two events can exist only if they have a timelike or lightlike separation. Thus, the temporal order of such events cannot be altered by a Lorentz transformation.[1] For causally related events it is therefore meaningful to talk of "earlier" or "later" in an absolute sense, i.e. independently of the inertial system. Only in this way is it possible to meaningfully introduce the concepts of cause and effect.

To each world point x there corresponds a light cone with its vertex at x. An event at x can be only effected by events in its past light cone, and it can have an effect only on events which lie in its future light cone.

4.2   Four-Vectors and -Tensors

As we already observed on page 63, because of the Lorentz transformations the coordinates in Minkowski space may themselves

---

[1] On the other hand, from causality one can deduce the Lorentz group. If one looks for the transformations which preserve the causal ordering of world points in Minkowski space (in the mathematical sense of a partial order), then one obtains, apart from a scale factor, precisely the Lorentz transformations. This interesting derivation of the Lorentz group can be found in the paper by E.C. Zeeman, J. Math. Phys. 5, 490 (1964).

be regarded as vectors. In accord with the discussion of tensors in Chapter 3, we now define four-vectors and four-tensors in Minkowski space in complete generality. We only have to identify the transformations (3.1) given above with the Lorentz transformations.

Thus, the four quantities

$$A^\mu = (A^0, \vec{A}), \quad \vec{A} = (A^1, A^2, A^3)$$

represent the contravariant components of a four-vector provided they transform under Lorentz transformations as the differentials of coordinates; since the Lorentz transformations are linear, this means that for the A^μ to be contravariant components of a vector, they ought to transform as the coordinates themselves.

The covariant components of a four-vector are obtained by contraction with the metric tensor,

$$A_\mu = g_{\mu\nu} A^\nu, \quad \text{i.e.} \quad A_\mu = (A^0, -\vec{A})$$

Then the scalar product of two vectors becomes

$$A^\mu B_\mu = g_{\mu\nu} A^\mu B^\nu = A^0 B^0 - \vec{A} \cdot \vec{B}$$

In conformity with the general definition (3.13), the contravariant (covariant) components of a four-tensor of rank r transform under Lorentz transformations as does the direct product of r contravariant (covariant) four-vectors.

In particular, for four-tensors of rank 2 one must distinguish between the following types of components:

$$T^{\mu\nu}, \quad T_{\mu\nu}, \quad T^\mu{}_\nu, \quad T_\mu{}^\nu$$

From the form of $g_{\mu\nu}$ it follows that, when raising or lowering a spatial index $(m,n=1,2,3)$[1] the sign of the component changes, but does not change when shifting the temporal idex. The nine spatial components T^{mn} form a three-dimensional tensor under spatial rotations. The components T^{0m} and T^{n0} behave under spatial rotations as three-dimensional vectors, and T^{00} as a scalar.

---

[1] While greek indices run from 0 to 3, roman indices will refer to the spatial components 1, 2, 3.

The trace of a second-rank tensor, T^μ_μ, is a four-scalar. The unit tensor is the mixed tensor δ^μ_ν (see p.69). Its trace is $\delta^\mu_\mu = 4$. The derivatives in the x^μ, compounded into the four-gradient

$$\frac{\partial\varphi}{\partial x^\mu} = \left(\frac{1}{c}\frac{\partial\varphi}{\partial t}, \vec{\nabla}\varphi\right), \quad \vec{\nabla} = \left(\frac{\partial}{\partial x^1}, \frac{\partial}{\partial x^2}, \frac{\partial}{\partial x^3}\right)$$

can be looked at as the components of a four-vector

$$\frac{\partial}{\partial x^\mu} \equiv \partial_\mu$$

since the differential of the scalar function $\varphi(x)$, i.e.

$$d\varphi(x) = \frac{\partial\varphi(x)}{\partial x^\mu} dx^\mu$$

is a scalar quantity. Then it follows that the divergence of a vector field $A^\mu(x)$,

$$\frac{\partial A^\mu}{\partial x^\mu} \equiv \partial_\mu A^\mu$$

is a scalar field. With the contravariant components of the differential operator

$$\partial^\mu \equiv \frac{\partial}{\partial x_\mu} = \left(\frac{1}{c}\frac{\partial}{\partial t}, -\vec{\nabla}\right)$$

the above inner product of two vectors may also be written in the form

$$\frac{\partial A_\mu}{\partial x_\mu} \equiv \partial^\mu A_\mu$$

If one forms the inner product of ∂_μ with itself, then one obtains the Lorentz invariant wave operator:

$$\frac{\partial^2}{\partial x^\mu \partial x_\mu} = \frac{1}{c^2}\frac{\partial^2}{\partial t^2} - \vec{\nabla}^2 \equiv \Box$$

With this operator the free-space wave equation for a massless scalar field $\varphi(x)$ or vector field $A^\mu(x)$ (such as the vector potential of electrodynamics) takes on the following form:

$$\Box \, \varphi \, (x) = 0 \quad , \quad \Box \, A^{\mu} \, (x) = 0$$

For a scalar field with mass m, the wave equation reads

$$(\Box + m^2 c^2) \, \phi \, (x) = 0$$

In Minkowski space one can perform integrations over a four-dimensional volume, a three-dimensional hypersurface, a two-dimensional surface, and along a curve in the space. Theorems analogous to those of Gauss and Stokes hold. The Gauss integral theorem permits the transformation of an integral over a closed hypersurface into an integral over the enclosed volume. In analogy to the three-dimensional case, this is achieved with the help of the four-vector ∂_{μ} via the substitution

$$d\sigma_{\mu} \longrightarrow d^4 x \, \frac{\partial}{\partial x^{\mu}}$$

Here $d^4 x = dx^0 dx^1 dx^2 dx^3$ stands for the four-dimensional volume element. Since the functional determinant of a Lorentz transformation is 1 (just as it is for a rotation), $d^4 x$ is a scalar. Hence the surface element $d\sigma_{\mu}$ is a four-vector. Its length equals the volume of the hypersurface element, and it is perpendicular to all directions which generate the hypersurface. Let us characterize a hypersurface by the parametric representation $x^{\mu} = x^{\mu}(u,v,w)$. In analogy to the vector flux $\vec{A} \cdot d\vec{f}$ of a three-dimensional vector $\vec{A}$ through the directed surface element

$$df_i = \varepsilon_{ijk} \frac{\partial x^j}{\partial u} \frac{\partial x^k}{\partial v} \, du \, dv$$

we must form, in the case of a four-vector $A^{\mu}(x)$, the volume of the parallelepiped spanned by A^{μ} and the tangent vectors $(\partial x^{\nu}/\partial u)du$, $(\partial x^{\rho}/\partial v)dv$, $(\partial x^{\kappa}/\partial w)dw$. Writing this quantity as the inner product of A^{μ} with the hypersurface element $d\sigma_{\mu}$ (vector flux), we get

$$\varepsilon_{\mu\nu\varsigma\varkappa} \, A^{\mu} \frac{\partial x^{\nu}}{\partial u} \frac{\partial x^{\varsigma}}{\partial v} \frac{\partial x^{\varkappa}}{\partial w} \, du \, dv \, dw = A^{\mu} d\sigma_{\mu} \quad ,$$

Here the hypersurface element is defined in the following manner:

$$d\sigma_\mu : = \varepsilon_{\mu\nu\varsigma\varkappa} \frac{\partial x^\nu}{\partial u} \frac{\partial x^\varsigma}{\partial v} \frac{\partial x^\varkappa}{\partial w} \, du \, dv \, dw$$

As an example, let us consider the most frequent case of a hypersurface given by x^0=const. It is advantageous to parametrize it by setting $u=x^1$, $v=x^2$, $w=x^3$. Since $\partial x^0/\partial x^m=0$, the definition of $d\sigma_\mu$ gives

$$d\sigma_\mu = (dx^1 \, dx^2 \, dx^3, 0, 0, 0)$$

This means: the projection of the hypersurface element onto the hyperplane x^0=const. is equal to the three-dimensional volume element, $d\sigma_0 = d^3x$. For a given instant x^0=const., $d\sigma_\mu$ is a timelike vector which, since $d^3x > 0$, points into the future. If one applies the above reasoning in context of the remaining three independent components of $d\sigma_\mu$, than one finds that the general hypersurface element has the representation

$$d\sigma_\mu = (dx^1 \, dx^2 \, dx^3, dx^0 \, dx^2 \, dx^3, dx^0 \, dx^1 \, dx^3, dx^0 \, dx^1 \, dx^2)$$

As a generalization of the Gauss theorem we then have the following rule for the integral over a vector A^μ:

$$\oint_\Sigma A^\mu \, d\sigma_\mu = \int_V d^4x \, \frac{\partial}{\partial x^\mu} A^\mu$$

4.3   The Full Lorentz Group

The transformation of a four-vector, especially that of x^μ, is given in Minkowski space by the expression

$$x'^\mu = L^\mu_{\ \nu} x^\nu$$

where $L^\mu_{\ \nu}$ stands for the matrix of the Lorentz transformation. Adapting the special Lorentz transformation (2.6) to this notation, i.e. writing

$$x'^0 = \gamma \left(x^0 - \frac{v}{c} x^1 \right)$$

$$x'^1 = \gamma \left(x^1 - \frac{v}{c} x^0 \right)$$

$$x'^2 = x^2$$

$$x'^3 = x^3$$

and using the definition of the equation (3.12)

$$L^\mu_{\ \nu} = \frac{\partial x'^\mu}{\partial x^\nu}$$

we can immediately read off the transformation matrix:

$$L^\mu_{\ \nu} = \begin{pmatrix} \gamma & -\frac{v}{c}\gamma & 0 & 0 \\ -\frac{v}{c}\gamma & \gamma & 0 & 0 \\ 0 & 0 & 1 & 0 \\ 0 & 0 & 0 & 1 \end{pmatrix} \tag{4.3}$$

Suppose now that the system K' is in motion relative to K not in the direction of x, but rather in the plane xy of K, and that the velocity vector $\vec{v}$ makes an angle Θ with the x axis. In order to find the corresponding Lorentz transformation between K and K' we can proceed as follows. First rotate the system K so that $\vec{x}$ becomes parallel to $\vec{v}$:

$$\overline{x} = R x$$

then transform into the moving system by using L^μ_ν (see equ.(4.3)):

$$\overline{\overline{x}} = L R x$$

and finally rotate back by the angle Θ:

$$x' = (R^{-1} L R) x = \Lambda x$$

We now convince ourselves that this procedure works as desired. The rotation in the xy plane with the angle Θ that we talked about is described by the matrix

$$
R = \begin{pmatrix}
1 & 0 & 0 & 0 \\
0 & \cos\theta & \sin\theta & 0 \\
0 & -\sin\theta & \cos\theta & 0 \\
0 & 0 & 0 & 1
\end{pmatrix}
$$

The associated inverse matrix R^{-1} corresponds to the rotation with the angle $-\theta$. Using $v_x = v\cos\theta$ and $v_y = v\sin\theta$, the multiplication results in

$$
R^{-1}LR = \begin{pmatrix}
\gamma & -\frac{v_x}{c}\gamma & -\frac{v_y}{c}\gamma & 0 \\
-\frac{v_x}{c}\gamma & 1+(\gamma-1)\frac{v_x^2}{v^2} & (\gamma-1)\frac{v_x v_y}{v^2} & 0 \\
-\frac{v_y}{c}\gamma & (\gamma-1)\frac{v_x v_y}{v^2} & 1+(\gamma-1)\frac{v_y^2}{v^2} & 0 \\
0 & 0 & 0 & 1
\end{pmatrix}
$$

This is the Lorentz transformation with $\vec{v}$ in the xy plane. The extension to the general case with $v_z \neq 0$ should be now evident, since all what one has to do is to extend the last row and column by terms which contain v_z:

$$
\Lambda^\mu_{\ \nu} = \begin{pmatrix}
\gamma & -\frac{v_x}{c}\gamma & -\frac{v_y}{c}\gamma & -\frac{v_z}{c}\gamma \\
-\frac{v_x}{c}\gamma & 1+(\gamma-1)\frac{v_x^2}{v^2} & (\gamma-1)\frac{v_x v_y}{v^2} & (\gamma-1)\frac{v_x v_z}{v^2} \\
-\frac{v_x}{c}\gamma & (\gamma-1)\frac{v_x v_y}{v^2} & 1+(\gamma-1)\frac{v_y^2}{v^2} & (\gamma-1)\frac{v_y v_z}{v^2} \\
-\frac{v_z}{c}\gamma & (\gamma-1)\frac{v_x v_z}{v^2} & (\gamma-1)\frac{v_y v_z}{v^2} & 1+(\gamma-1)\frac{v_z^2}{v^2}
\end{pmatrix}
$$

If one applies the transformation $\Lambda^\mu_{\ \nu}$ onto the vector $x^\mu = (x^0, \vec{x})$, one obtains, in accord with the earlier result (2.8),

$$\vec{x}' = \vec{x} + (\gamma - 1)\frac{(\vec{x}\cdot\vec{v})\vec{v}}{v^2} - \gamma\,\vec{v}\,t$$

$$t' = \gamma\,(t - \frac{\vec{x}\cdot\vec{v}}{c^2})$$

If one complements this Lorentz transformation in an arbitrary direction of the velocity $\vec{v}$ (3 parameters) with spatial rotations (3 parameters) and translations in space and time (4 parameters), then one arrives at the Poincaré group (also called the inhomogeneous Lorentz group):

$$x'^\mu = \Lambda^\mu{}_\nu x^\nu + a^\mu$$

If no translations occur ($a^\mu = \{0\}$), then one speaks of the homogeneous Lorentz group.

The matrix Λ of the homogeneous Lorentz transformation has the following property:

$$\Lambda^\varkappa{}_\mu\, g_{\varkappa\lambda}\, \Lambda^\lambda{}_\nu = g_{\mu\nu} \tag{4.4}$$

This comes about because for arbitrary vectors x and y we have the scalar product

$$\Lambda^\varkappa{}_\mu x^\mu\, g_{\varkappa\lambda}\, \Lambda^\lambda{}_\nu y^\nu = g_{\mu\nu} x^\mu y^\nu$$

Using the definition $(\Lambda^T)^\mu{}_\nu = \Lambda_\nu{}^\mu$ of the transposed matrix, this can be rewritten as

$$(\Lambda^T)_\mu{}^\varkappa\, (g\Lambda)_{\varkappa\nu} = g_{\mu\nu} \tag{4.5}$$

These equations will prove useful in the subsequent discussion of the four pieces of the Lorentz group.

To begin with, let us observe that so far the discrete operations have not been yet taken into account. But it is clear that the operations of spatial and and temporal reflection, i.e.

$$x'^\mu = \Pi^\mu{}_\nu x^\nu, \qquad x'^\mu = \theta^\mu{}_\nu x^\nu$$

with

$$\Pi^{\mu}_{\ \nu} = \begin{pmatrix} 1 & & & 0 \\ & -1 & & \\ & & -1 & \\ 0 & & & -1 \end{pmatrix} \quad , \quad \theta^{\mu}_{\ \nu} = \begin{pmatrix} -1 & & & 0 \\ & 1 & & \\ & & 1 & \\ 0 & & & 1 \end{pmatrix}$$

leave the scalar product $x \cdot y$ invariant, and therefore they belong to the full Lorentz group. However, they cannot be continuously connected to the identity element **1**. Therefore we expect that the full Lorentz group consists of disconnected pieces, each being in itself a connected component. We now proceed to find the four pieces of the Lorentz group.

Taking the determinants of both sides of equation (4.5) we get

$$|\Lambda^T| \cdot |g| \cdot |\Lambda| = |g|$$

$$|\Lambda| = \pm 1$$

The transformations continuously connected to the unit element **1** must obey the condition $|\Lambda| = +1$, because the determinant of the unit matrix is +1. The discrete operations Π and Θ have determinants -1, while the product $\Pi\Theta$ (total inversion) has the determinant +1. This immediately leads to the distinguishing of two groups, L_+ and L_-. So as to obtain further subdivision, one writes down the oo-component of equation (4.4),

$$g_{00} = 1 = \Lambda^{\varkappa}_{\ 0} \, g_{\varkappa\lambda} \Lambda^{\lambda}_{\ 0} = (\Lambda^0_{\ 0})^2 - \sum_{k=1}^{3} (\Lambda^k_{\ 0})^2$$

Therefrom follows the condition

$$(\Lambda^0_{\ 0})^2 \geq 1$$

which leads to a further subdivision of the Lorentz group into the pieces with

$$\Lambda^0_{\ 0} \geq 1 \quad \text{and} \quad \Lambda^0_{\ 0} \leq -1$$

The transformations continuously connected to the unit element clearly satisfy the condition $\Lambda^o{}_o > 1$, but for the total inversion $\Pi\Theta$ we have $\Lambda^o{}_o = -1$.

Accordingly, we arrive at the following decomposition of the Lorentz group into four pieces:

$$L_+^\uparrow \; : \; |\Lambda| = +1, \; \text{sign } \Lambda^o{}_o = +1 \quad \text{contains} \quad \mathbf{1}$$

$$L_-^\uparrow \; : \; |\Lambda| = -1, \; \text{sign } \Lambda^o{}_o = +1 \quad " \quad \Pi$$

$$L_+^\downarrow \; : \; |\Lambda| = +1, \; \text{sign } \Lambda^o{}_o = -1 \quad " \quad \Pi\,\Theta$$

$$L_-^\downarrow \; : \; |\Lambda| = -1 \; \text{ sign } \Lambda^o{}_o = -1 \quad " \quad \Theta$$

This result is visualized in Fig.4.3. One can combine two pieces in several ways, so as to obtain several interesting subgroups of the full Lorentz group. These are indicated in the figure.

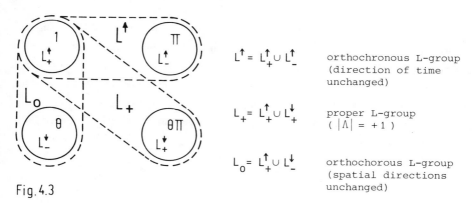

$L^\uparrow = L_+^\uparrow \cup L_-^\uparrow$ orthochronous L-group (direction of time unchanged)

$L_+ = L_+^\uparrow \cup L_+^\downarrow$ proper L-group ($|\Lambda| = +1$)

$L_o = L_+^\uparrow \cup L_-^\downarrow$ orthochorous L-group (spatial directions unchanged)

Fig.4.3

For the formulation of relativistic quantum field theory, the above determined structure of the Poincaré group and its properties as a Lie group play a fundamental role. In particular, so as to satisfy the requirement of invariance, the relativistic states of a system are introduced as elements of the representation space of the unitary representations of this symmetry group.[1] Besides the tensors we discussed in the foregoing spinor representations also occur, which describe particles (or fields) with half integral spin. – For our purposes it suffices to pay attention only to the proper orthochronous Lorentz group $L_+^\uparrow$, or rather to the corresponding Poincaré group $P_+^\uparrow$.

4.4   Geometrical Representation of the Lorentz Transformation

The Lorentz transformations may be interpreted as rotations in Minkowski space. Apart from ordinary rotations in the three spatial planes, one may have "rotations" in the planes tx, ty, tz. We explain this on hand of the special Lorentz transformation (2.6),

$$x' = \gamma\,(x - vt)$$

$$t' = \gamma\,(t - \frac{v}{c^2}x)$$

which can be interpreted as a "rotation" in the tx plane. One must observe that in this case the squared separation

$$(ct')^2 - x'^2 = (ct)^2 - x^2$$

stays invariant. Since

$$\cosh^2 \alpha - \sinh^2 \alpha = 1$$

one can arrange this to hold with the use of the hyperbolic functions

$$\sinh \alpha = \frac{e^\alpha - e^{-\alpha}}{2} \quad , \quad \cosh \alpha = \frac{e^\alpha + e^{-\alpha}}{2}$$

by simply putting for the rotation

---

[1] The reader is referred to the relevant textbooks on the theory of elementary particles or field theory. Regarding an introduction to the representation theory of the Poincaré group see, for example, L. Fonda and G.C. Ghirardi: Symmetry Principles in Quantum Physics (Marcel Decker, New York, 1970); Wu-Ki Tung: Group Theory in Physics (World Scientific, Singapore, 1985). Here one also can find further references to the more detailed literature.

$$x' = x \cosh \alpha - ct \sinh \alpha$$

$$ct' = ct \cosh \alpha - x \sinh \alpha$$

The angle α is determined by the relative velocity v. Comparing the two ways of writing the transformation it follows that

$$\tgh \alpha = \frac{v}{c}$$

Now we recall the connection between the hyperbolic and the trigonometric functions,

$$\sinh \alpha = -i \sin(i\alpha), \quad \cosh \alpha = \cos(i\alpha)$$

Thus, the special Lorentz transformation is described by a rotation in the complex plain x,ict with the imaginary angle iα, where

$$\alpha = \operatorname{Artgh} \frac{v}{c}$$

Consequently, it is plausible to introduce the time coordinate as the fourth component of a four-vector, $x_4 = ict$, and to use Euclidean metric in this space. For the squared separation defined on page 35 we then have

$$-s^2 = x_1^2 + x_2^2 + x_3^2 + x_4^2 = x^2 + y^2 + z^2 - c^2 t^2$$

Nevertheless, we will continue to use the notational system so far used which is based on the pseudo-Euclidean metric tensor (4.1). –

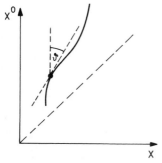

Fig.4.4

The interpretation of Lorentz transformations as rotations in the coordinates x,y,z,ict stems from Poincaré.

We now enlarge upon the geometric visualization of the Lorentz transformation in a space-time diagram. Fig.4.4 shows the world line of a non-zero-mass particle in the xx^0 plane, and for comparison also the world line of a light wave passing through 0 (the photon mass is zero). The latter one is precisely the bisector if the axes x^o and x are equipped with the same units. In contrast, the tangent of the world line of the m≠0 particle makes, at any point, an angle ϑ with the time axis which is always less than 45°. This is so because

$$\frac{dx}{dx^o} = \frac{v}{c} < 1$$

The transition $K \rightarrow K'$ to another inertial system according to the special Lorentz transformation L (see equ.(2.6)) is illustrated in Fig.4.5. From the condition $x'^o=0$ that characterizes points on the x' axis it follows that $x^o=(v/c)x$, which means that the coordinates x' lie in the xx^o plane on a straight line through 0 with slope

$$tg \; \varphi = \frac{v}{c}$$

The corresponding inclination toward the x axis is $< 45^o$. From the condition x'=0 that characterizes the x'^o axis it follows that $x^o=(c/v)x$. In respect to the bisector, this straight line is mirror symmetric to the x' axis.

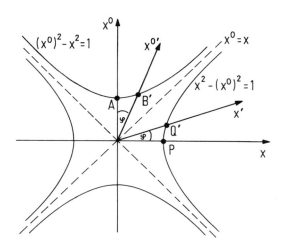

Fig.4.5

This must be so because in K' the world line of light (of photons) is again the bisector of the coordinate axes, on account of the invariance of lightlike separations. Observe that the new axes of the coordinate system K' are no longer orthogonal.

In this manner, the relative nature of simultaneity becomes directly apparent. Events on the x' axis are simultaneous for the observer K' (they have $x'^o = 0$), whereas for the observer K they occur in succession.

The units of meter sticks in K and K' are determined by gauge curves, for which we take here the branches of two hyperbolae, viz. $(x^o)^2-(x)^2=1$ in the timelike and $(x)^2-(x^o)^2=1$ in the spacelike region (see Fig.4.5). For rotations in the Euclidean plane the gauge curve is the unit circle $x^2+y^2=1$. The units on the axes of K' are determined by their intersections with the gauge hyperbolae. Now we can read off time dilation and length contraction from the graphic representation (see Fig.4.6 and 4.7).

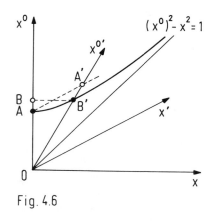

Fig. 4.6

a) When a clock at rest in K' shows precisely the unit of time valid in this system (world point B' on the x'^o axis), then in system K one measures more than one time unit for this event, since $\overline{OB} > \overline{OA}$. Now let the proper time unit valid in K be indicated by the event A. The coordinates of A are determined in K' by the corresponding parallel projections. One sees that in K', more than one time unit has passed between the two events at O and A', because $\overline{OA'} > \overline{OB'}$. In summary: the phenomenon of mutual time dilation is an expression for projections upon the coordinate axes of the reference frames in relative motion.

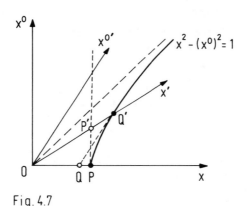

Fig. 4.7

b) In an analogous manner the length contraction may also be geometrically interpreted. Let a meter stick of length 1 be at rest in K'. The world line of its endpoint Q' (see Fig.4.7) runs parallel to x'^0 and intersects the x axis at Q. In order to determine the length of the meter stick in K one must measure simultaneously the coordinates of its endpoints. These are given by the intersection of a straight line parallel to x (i.e., x^0=const.) with the world lines of the endpoints. For the instant $x^0 = 0$, this means exactly the distance $\overline{OQ}$. But this is shorter than the length unit $\overline{OP}$ in K. Conversely, when the unit stick $\overline{OP}$ is at rest in K, then the world line from P runs parallel to the x^0 axis and intersects the x' axis at P'. In such manner, the length of the stick at the time $x'^0 = 0$ is given for the observer in K' by the length $\overline{OP'}$, which is shorter than the unit length $\overline{OQ'}$ in K'.

4.5   Proper Time, Velocity, Acceleration

The motion of a mass point is described in Minkowski space by the world line $x^\mu = x^\mu(\lambda)$, where λ is a parameter yet to be determined. It should be useful to consider the world line as function of an invariant time parameter, because when one introduces such concepts as velocity or acceleration, one will have to take the time derivative of the world point x^μ. The time parameter used ought to be an invariant, otherwise the time derivative of x^μ would not yield a four-vector. The ordinary time coordinate t does not satisfy this condition, since we know that it transforms as the zero-component of the vector x^μ.

With the invariant separation ds (see equ.(4.2)) and the likewise invariant light velocity c one can define the invariant time interval as

$$d\tau = \frac{1}{c} \sqrt{g_{\mu\nu} \, dx^\mu \, dx^\nu} = dt \sqrt{1 - \frac{v^2}{c^2}} \qquad (4.6)$$

Here v(t) is the velocity of the mass point. This equation expresses $d\tau$ by the time t of that reference system K which was chosen for the description of the process. The invariant time interval $d\tau$ may be calculated in any arbitrary inertial system. The physical meaning of $d\tau$ becomes clear if one passes to the rest system of the mass point. In this system v=0, hence $d\tau = dt$. This means that a clock attached to the mass point, i.e., a comoving clock, shows precisely the time interval $d\tau$. Thus, the quantity $d\tau$ which was introduced on account of invariance considerations, coincides with the proper time which we met already in the discussion of time dilation and the twin paradox.

The finite time interval shown by the comoving clock will now be given by

$$\Delta\tau = \int_A^B dt \sqrt{1 - \frac{v^2}{c^2}}$$

This integral depends not only on A and B but also on the world line connecting these world points. The path dependence of $\Delta\tau$ becomes important in the clarification of the twin paradox, since a clock shows the time which equals the integral of the proper time interval along its world line. For a clock at rest the world line is a straight line parallel to the x^0 axis; for a moving clock which returns to its place of departure, the world line is curved (see p.51, Fig.2.5). As we saw earlier, the time interval along the straight world line (clock at rest) is always greater as the one taken along the curved world line (moving clock). Thus, the above integral has maximal value if the integration path connecting the world points is a straight world line. This property arises from the pseudo-Euclidean metric of Minkowski space. In contrast, the analogous integral of the line element in an Euclidean space, taken along a straight line, has a minimum value, because the straight line is the shortest connection between two points.

Now we define the velocity four-vector as the derivative of x^μ with respect to the proper time,

$$u^\nu = \frac{dx^\nu}{d\tau} \qquad (4.7)$$

Since $dt = \gamma d\tau$, this implies for the individual components

$$u^0 = c\gamma \quad , \qquad u^1 = \frac{dx^1}{dt}\frac{dt}{d\tau} = v_x \gamma \quad ,$$

$$u^2 = v_y \gamma \quad , \qquad u^3 = v_z \gamma \qquad (4.8)$$

The inner product of this four-vector with itself gives

$$g_{\mu\nu} u^\mu u^\nu = (u^0)^2 - \vec{u}^2 = c^2 \qquad (4.9)$$

Thus, the four-velocity is a timelike vector with magnitude c. Its direction is given by the tangent to the world line at each point x^μ. Equation (4.9) reflects the invariance of the speed of light c.

In a similar manner one defines the four-acceleration as the derivative of the four-velocity with respect to the proper time,

$$\dot{u}^\nu = \frac{du^\nu}{d\tau} = \frac{d^2x^\nu}{d\tau^2} \qquad (4.10)$$

If $\dot{u}^\nu \neq 0$, then the velocity $\vec{v}$ is no longer constant in time, $\vec{b}=d\vec{v}/dt$, and one finds for the components of the acceleration

$$\dot{u}^0 = \frac{d}{dt}\left(\frac{c}{\sqrt{1-\beta^2(t)}}\right)\frac{dt}{d\tau} = \frac{\vec{v}\cdot\vec{b}}{c\,(1-\beta^2)^2}$$

$$\dot{u}^1 = \frac{d}{dt}\left(\frac{v_x}{\sqrt{1-\beta^2(t)}}\right)\frac{dt}{d\tau} = \frac{b_x}{1-\beta^2} + \frac{v_x\,(\vec{v}\cdot\vec{b})}{c^2\,(1-\beta^2)^2}$$

with corresponding expressions for $\dot{u}^2$ and $\dot{u}^3$. The inner product gives (because $v/c < 1$)

$$g_{\mu\nu}\,\dot{u}^\mu\dot{u}^\nu = -\frac{(\vec{b})^2 - \left[\frac{\vec{v}}{c}\times\vec{b}\right]^2}{(1-\beta^2)^3} < 0$$

Thus, the four-acceleration is a spacelike vector. Forming the time derivative of equation (4.9), one gets

$$g_{\mu\nu}\,u^\mu\dot{u}^\nu = 0 \tag{4.11}$$

which means that the four-vector of the velocity is orthogonal to that of the acceleration.

Chapter 5

RELATIVISTIC MECHANICS

Newtonian mechanics obeys the Galilean relativity principle and hence, at high velocities it leads to false predictions. Therefore, mechanics must be so formulated that it conforms to the universal Einstein relativity principle (see Chapter 2). This relativistic mechanics ought to contain Newtonian mechanics as a limiting case for small velocities ($v \ll c$). Einstein's relativity principle can be fulfilled in a simple manner by writing the equations of motion of mechanics as covariant equations in Minkowski space. The changes from the Newtonian mechanics which arise in this way, can be verified experimentally.

5.1   Dynamical Equation for a Mass Point

In this section we introduce the entities of relativistic mass point mechanics as generalizations of the corresponding Newtonian quantities. One arrives at the relativistic formulation in a more systematic manner via the invariant action integral. This will be discussed in Chapter 5.5.

To start with, we observe that the Newtonian law of inertia (Axiom 1) is invariant under Lorentz transformations. For if a particle moves without acceleration in system K, then it remains acceleration-less if one passes to the transformed system K'. In Newtonian mechanics the law of inertia is expressed by the equations

$$m \vec{v} = \text{const} \quad , \quad \text{or} \quad \frac{d}{dt} (m \vec{v}) = 0 \tag{5.1}$$

Correspondingly, we shall describe this law relativistically by the constancy of the four-momentum:

$$p^\nu = \text{const} \quad , \quad \text{or} \quad \frac{d}{d\tau} p^\nu = 0 \tag{5.2}$$

Here τ stands for the proper time. The inertia of the particle is described by an invariant quantity, the mass m of the particle, more properly called its rest mass. Functioning as a characteristic quantity of the particle, it has the same value in all reference frames. Hence, it is sufficient to determine it in the rest system or at small velocities. Proceeding, we define the four-momentum p^ν analogously to Newtonian mechanics (since this holds for $v \ll c$), i.e., we express p^ν in terms of the four-velocity:

$$p^\nu = m \, u^\nu = \left(\frac{mc}{\sqrt{1-\beta^2}} \, , \, \frac{m\vec{v}}{\sqrt{1-\beta^2}} \right) \tag{5.3}$$

The spatial components of the momentum p^ν become in the limiting case $v \ll c$ the components of the customary momentum. Equations (5.1) express conservation of momentum for closed systems. Momentum conservation is the consequence of invariance against spatial translations, i.e. of the homogeneity of space. Similarly, the conservation of the four-momentum (5.2) can be looked upon as the consequence of translation invariance in Minkowski space, that is, homogeneity of space and time. The quantity which is conserved for a free particle on account of invariance under temporal translations, is just the energy. Therefore we expect that the particle energy is proportional to the temporal component p^o of the four-momentum.

The above definition of the four-momentum suggests the following relativistic generalization of Newton's law of motion (Axiom 2):

$$\frac{d}{d\tau} p^\nu = m \frac{du^\nu}{d\tau} = K^\nu \tag{5.4}$$

This covariant form of the equation of motion was first given by Minkowski.[1] Therefore the four-vector K^ν is called the Minkowski force. In order to compare with the ordinary force $\vec{F}$ that occurs in the Newtonian law

$$\frac{d}{dt}(m\vec{v}) = \vec{F}$$

we consider the spatial components in equation (5.4):

$$\frac{1}{\sqrt{1-\beta^2}}\frac{d}{dt}\frac{m\vec{v}}{\sqrt{1-\beta^2}} = \vec{K}$$

This relation may be cast into the form of the Newtonian equations,

$$\frac{d}{dt}\frac{m\vec{v}}{\sqrt{1-\beta^2}} = \vec{K}\sqrt{1-\beta^2} \tag{5.5}$$

Here, on the left-hand side, we have the ordinary time derivative of the Newtonian momentum $m_r\vec{v}$ with a velocity dependent mass,

$$m_r = \frac{m}{\sqrt{1-\beta^2}} \tag{5.6}$$

Therefore, on the right-hand side, we must have the ordinary acting force:

$$\frac{d}{dt}m_r\vec{v} = \vec{F} \tag{5.7}$$

Comparing this with equation (5.5), we see that the connection between the Minkowski force $\vec{K}$ and the ordinary force $\vec{F}$ is given by

$$\vec{K} = \frac{\vec{F}}{\sqrt{1-\beta^2}} \tag{5.8}$$

The mass m_r defined in equation (5.6), also called dynamic mass or momentum mass, is not an invariant quantity. It depends on the velocity of the particle relative to the frame of reference and for $v = 0$ it

---

[1] The rest mass m is a constant, provided the particle does not decay. This has been assumed in equ. (5.4).

becomes the rest mass.[1] Its significance is that it can be measured directly. Actually, the relation (5.6) has been experimentally verified with an accuracy of about 0.05% (see footnote on p.31). For $v \to c$ the relativistic mass m_r becomes infinite. Therefore it is impossible to accelerate a particle with rest mass $m \neq 0$ in a finite time span to reach light velocity. This is so because for $v \to c$ it follows from the integrated equation of motion in its Newtonian form (5.7) that

$$\int \vec{F} \, dt = \frac{m \, \vec{v}}{\sqrt{1 - \frac{v^2}{c^2}}} \to \infty$$

But for $m \neq 0$ and a finite force this cannot be fulfilled in a finite time interval. In this way we can understand the experiment regarding the limiting velocity, which was discussed in Chapter 2.2. However, the relativistic variability of mass is not a dynamical effect but rather a result of time dilation which comes to play when one passes from the ordinary velocity to the four-velocity.

In this context, the following comment may prove useful. The spatial components of a four-vector form, by definition, a spatial (3-dimensional) vector. (Examples: $\vec{u}$, $\vec{K}$, etc.). On the other hand, the components of an ordinary vector do not necessarily transform as the spatial components of a four-vector. One may multiply the components of an ordinary vector by a function of $\beta = v/c$ without altering the behaviour under spatial rotations. But such a factor makes an essential difference regarding the behaviour of these components under Lorentz transformations. Thus, if one wants to construct from an ordinary vector the spatial components of the corresponding four-vector, one must find the appropriate factor. For the vector $\vec{F}$ we found that the correct factor is γ (see equ.(5.8)). To give another example, one obtains in the same manner the spatial components of the four-velocity from the ordinary velocity, $\vec{u} = \gamma \vec{v}$.

We now ask: what is the meaning of the temporal component of the Minkowski force K^ν. To find out, we contract the vectors in

---

[1] For the invariant description of the inertial behaviour of a particle, only its rest mass m can be used, because it has the same value in all reference systems.

equation (5.4) with the velocity u_ν. Since velocity and acceleration are orthogonal (see equ.(4.11)), it follows that

$$K^\nu u_\nu = 0$$

and from this

$$K^0 = \frac{\vec{K} \cdot \vec{u}}{u_0} = \gamma \frac{\vec{F} \cdot \vec{v}}{c}$$

This gives for the temporal components in equation (5.4) the relation

$$\gamma \frac{d}{dt} \frac{mc}{\sqrt{1-\beta^2}} = K^0 = \gamma \frac{\vec{F} \cdot \vec{v}}{c}$$

or

$$\frac{d}{dt} \frac{mc^2}{\sqrt{1-\beta^2}} = \vec{F} \cdot \vec{v} \qquad (5.9)$$

But $\vec{F} \cdot \vec{v}$ means the work done on the particle by the force $\vec{F}$ during a unit time interval. Hence, the left-hand side of equation (5.9) must represent the corresponding temporal change of the energy. Consequently,

$$E = \frac{mc^2}{\sqrt{1-\beta^2}} \qquad (5.10)$$

must be interpreted as the energy of the particle (point mass). The temporal change of the energy equals the power of the external force. However, this determination of energy expression is formally not unique. Since on the left-hand side of equation (5.9) we have the time derivative, the equation would also be satisfied by

$$\tilde{E} = \frac{mc^2}{\sqrt{1-\beta^2}} + \text{const}$$

In the next Chapter we shall convince ourselves that one must choose the additive constant zero, and hence the energy is uniquely determined by equation (5.10).

5.2   Momentum, Energy, Mass

In this Chapter we will discuss the quantities just introduced in more detail. Equation(5.10) and the relation $p^o = \gamma mc$ (cf.equ.(5.3)) determines the proportionality factor in the relation expected between the energy and p^o:

$$E = c\, p^o$$

The temporal component of p^ν coincides with the particle energy up to the factor $1/c$. Therefore, it is customary to talk of the energy-momentum vector,

$$p^\nu = \left(\frac{E}{c}, \vec{p}\right) \; ; \; \vec{p} = \gamma m \vec{v} \qquad (5.11)$$

Lorentz transformations mix the energy and the components of momentum,

$$p'^\mu = \Lambda^\mu{}_\nu p^\nu$$

Thus, energy and momentum are no longer separate covariant quantities as they are under spatial rotations. Since $m > 0$, p^ν is a timelike vector,

$$g_{\mu\nu}\, p^\mu p^\nu = m^2 c^2$$

This relation reflects the invariance of the rest mass m. The equation also implies the relativistic connection between energy and momentum,[1]

$$E = c\sqrt{\vec{p}^2 + m^2 c^2} \qquad (5.12)$$

which is used in many applications of the theory. If the particle has zero rest mass (e.g., a photon), then p^ν is a lightlike vector and equation (5.12) gives

$$|\vec{p}| = \frac{E}{c} \qquad (5.13)$$

---

[1] So as to simplify writing, one often uses units such that $c = 1$.

Such particles move with the speed of light, because eliminating the mass from $\vec{p}=\gamma m\vec{v}$ and $E=\gamma mc^2$, one is led to

$$\vec{p} = \frac{E\,\vec{v}}{c^2}$$

This is compatible with equation (5.13) only if $|\vec{v}|=c$. In case of $m \neq 0$, equation (5.13) may still be used as an approximation, at least in the extreme relativistic case when E is much larger than mc^2.

For a particle with nonvanishing rest mass, the energy becomes infinite when $v \rightarrow c$. This again illustrates that such particles cannot be given the velocity of light. But for a particle with zero rest mass equation (5.10) leads for $v \rightarrow c$ to the undefined expression 0:0, so that the energy can remain finite. But then, the velocity of such a particle must always be c. According to equation (5.13), the energy will then be $E=c|\vec{p}|$.

When one solves the equation

$$p_\mu\, p^\mu = \left(\frac{E}{c}\right)^2 - \vec{p}^{\,2} = m^2\, c^2$$

for E, in principle one can choose either sign of the square root,

$$E = \pm\, c\, \sqrt{\vec{p}^{\,2} + m^2\, c^2}$$

which means that unphysical negative energy values can occur. For noninteracting particles one simply could exclude the negative sign. But if one tries to set up systematically an equation analogous to the Schrödinger equation which would conform to the relativistic relation between energy and momentum, then the complete system of solutions of such an equation unavoidably contains solutions with a "negative energy". One can interpret this doubling of solutions by introducing the notion of antiparticles, which have a charge opposite to that of the particle. Thus, in 1928 P.A.M. Dirac was led to the positron as the hypothetical antiparticle of the electron. The experimental discovery of the positron in 1932 by C.D. Anderson verified the existence of antiparticles.

The relativistic wave equation given in Chapter 4.2 may be derived from the relation

$$\left(\frac{E}{c}\right)^2 - \vec{p}^2 = m^2 c^2$$

if one employs the quantum mechanical correspondence (with units choosen such that $h/2\pi = 1$)

$$\vec{p} \rightarrow -i\vec{\nabla} \, , \quad E \rightarrow i\frac{\partial}{\partial t}$$

beween numbers and operators. Putting in into the former equation these corresponding operators, one obtains the Klein–Gordon equation

$$(\Box + m^2 c^2)\, \Phi(x) = 0$$

which we already met on p.90. Here

$$\Box := \frac{1}{c^2}\frac{\partial^2}{\partial t^2} - \vec{\nabla}^2$$

acts on the scalar field $\Phi(x)$. This equation is covariant, i.e. it holds in the same form for all inertial systems. Its complete set of solutions contains such that have negative energy.

We return to equation (5.10). One first observes that, in consequence of this relation, the energy of a free particle with rest mass $m \neq 0$ does not vanish for $v=0$ in relativistic mechanics, but rather, it takes on the rest energy value

$$E_0 = m c^2 \qquad\qquad (5.14)$$

For small velocities we have

$$E = m c^2 \left(1 + \frac{1}{2}\frac{v^2}{c^2} - \dots\right) \approx m c^2 + \frac{1}{2} m v^2$$

that is, in this approximation the rest energy is added to the nonrelativistic kinetic energy term. The quantity E is therefore also called the total energy of the particle. But one must be aware that E does not contain the potential energy in an external field when such a field acts on the particle.

Since the energy was identified in equation (5.10) on the basis of its temporal change, this argumentation determines E only up to an additive constant. One is then tempted to choose for this constant the negative value of the rest energy. Then $\tilde{E}=E-mc^2$ would go over into the kinetic energy of Newtonian mechanics when $\beta \ll 1$. However, one must not expect, and certainly must not demand that all quantities of relativistic mechanics take on their classical (Newtonian) form for $\beta \ll 1$. On the other hand one does have the firm criterion that for $\beta \ll 1$ the Lorentz transformation must go over into the Galilei transformation and that, therefore, for $\beta \ll 1$ the customary velocity addition law must hold. Now, in order that for $v \ll c$ the transformation

$$p'^1 = \gamma \left(p^1 - v \frac{E}{c^2} \right)$$

should yield the customary velocity addition law, one must surely have $E/c^2 \to m$ for $\beta \to 0$. Then, in the nonrelativistic limit

$$p'^1 = p^1 - mv , \quad \text{or} \quad v'^1 = v^1 - v$$

as desired. In contrast, if one made the tempting choice $\tilde{E}=E-mc^2$ for the expression of the energy, this limiting behaviour would not hold.[1] Hence the additive constant must be set zero and E is uniquely determined by the relation (5.10).

In this manner, the theory of relativity leads to the important conclusion that the energy of a particle (with $m \neq 0$) at rest is mc^2. The relations

$$E_0 = m c^2$$

$$E = \frac{m c^2}{\sqrt{1 - \beta^2}} \tag{5.15}$$

for a particle at rest and a particle in motion, respectively, are the famous Einstein formulae. They express the equivalence of mass

---

[1] One should also take it as a warning that $\tilde{E}$ would be the difference of two terms with different transformation properties.

and energy. According to these formulae, these two physical entities are proportional to each other. Since, after all, the relations (5.15) follow from Lorentz invariance, they are universally valid beyond the realm of mechanics and thus have a very deep significance. This statement (5.15) is often called as the universal equivalence principle of mass and energy and it was formulated in this generality by Einstein.[1] According to this, there does not exist a mass without energy (or vice versa) and every change in energy is connected with a corresponding change of inertial mass. The relations (5.15) have many experimental verifications. So far, no deviations could be found.

For a closed system (i.e. if no external forces act on the particle) energy–momentum conservation holds true. But from the mass-energy relation it follows that, if E=const., then also the relativistic mass m_r remains constant,

$$m_r = \frac{m}{\sqrt{1 - \beta^2}} = const$$

While in Newtonian mechanics energy conservation and conservation of mass are two different laws, in relativity theory there is only one conservation law, the conservation of energy or, in accord with the above relation, the conservation of relativistic mass m_r. Apparently, the rest mass m need not be conserved. This indicates that the invariant mass m is basically different from the mass that appears in Newtonian mechanics. The nonconservation of rest mass is immediately clear from the fact that amongst the much observed transmutation processes of particles one also finds such where particles with $m \neq 0$ go over into particles with rest mass zero. Examples are the decays $\pi^0 \to 2\gamma$ or $e^+ + e^- \to 2\gamma$. Nevertheless, because of its relativistic

---

[1] Einstein considered this equivalence as a particularly important consequence of the theory of special relativity and he proposed several proofs for it. We note as a curiosity that his first derivation in the famous work of 1905, Ann. Phys. 18, 639 (1905) was criticized later because of logical inconsistency. This was pointed out by H.E. Ives, J. Opt. Soc. America 42, 540 (1952). But the significance of Einstein's contribution is in no way diminished by this criticism.See also the discussion on this topic in the historical review on E=mc^2 provided recently by L.Fadner,Am.J.Phys.56,114(1988).

invariance, rest mass is an important characteristic of material objects (particles). In contrast, the "momentum mass" m_r depends on the particle's state of motion, and it transforms as the energy E, i.e. as the temporal component of a four-vector.

So as to obtain the kinetic energy of a particle, one must subtract the rest energy E_0 from the total energy, i.e.

$$E_{kin} = E - mc^2 = mc^2 \left[\frac{1}{\sqrt{1-\beta^2}} - 1 \right] \qquad (5.16)$$

From this we find the relativistic relation between v^2 and the kinetic energy:

$$v^2 = c^2 \left[1 - \frac{1}{\left[1 + \frac{E_{kin}}{mc^2}\right]^2} \right] \qquad (5.17)$$

Experiments verify this relation. This formula reproduces the curve in Fig.2.1 (page 16) which was obtained in the experiment probing the limiting velocity. In the nonrelativistic limit and with $E_{kin} \ll mc^2$ it leads to the usual expression $v^2 = 2E_{kin}/m$.

As we already mentioned, the Einsteinian relations (5.15) have a universal validity. In their derivation there was no assumption concerning the elementarity or compositness of the particle. Therefore the relations (5.15) are also applicable to a body composed of many particles or to composite systems. Then v stands for the velocity of the entire body (system). If the total rest mass of the body is denoted by M, then for a body at rest the rest energy is given by

$$E_0 = Mc^2$$

Since M > 0, the rest energy of the closed system is positive, just as for a single particle.

The rest energy Mc^2 of a composite body (system) contains the rest energies of the particles that make it up, and it contains also the kinetic energy of these particles, as well as the energy of their interactions. In other words, the rest energy contains the entire internal energy of the system, i.e., the thermal energy, excitation energy, binding energy, and so on. If one increases the internal energy, then also its rest mass will increase. Thus, an excited atom is heavier than the same atom in its groundstate.

In summary, the sum $\sum_i m_i c^2$ does not yield the rest energy Mc^2 of the body. Consequently, the rest mass of a composite system is surely not equal to the total rest mass of its components. This again shows that there is no conservation in relativity theory for the rest mass. On the other hand, we have the conservation of energy-momentum which includes also the rest energy or rest mass of the particles. Therefore, provided this conservation law allows it and provided all other possible conservation laws are also fulfilled, rest mass (energy) can be transformed into the equivalent energy (rest mass) which, according to equation (5.15) corresponds to it. This is best illustrated by the creation and annihilation of elementary particles which we will consider again later on.

When a stable bound system is formed from previously free particles (fusion), energy is liberated. It is necessary to invest again at least this amount of binding energy if one wants to decompose the system into its constituents. Thus, the rest mass of a stable system is always less than the sum of the rest masses of the individual particles,

$$M < \sum_i m_i$$

In that case, a spontaneous decay of the system into particles with the masses m_i is not possible. The difference of masses

$$\Delta M = \sum_i m_i - M \qquad (5.18)$$

which, for stable systems is a positive quantity, is called the mass defect. The equivalent amount of energy is nothing but the already mentioned binding energy

$$E_B = \Delta M c^2$$

Thus, the mass defect that occurs for bound systems may be interpreted in terms of the binding energy. Both entities are a measure of the stability of matter and therefore they are of crucial importance, especially in atomic and nuclear physics.

On th other hand, a system may decompose into subsystems (i.e., it may be unstable) whenever the sum of the rest masses of the sub-systems is smaller then the rest mass of the entire system,

$$M > \sum_i M_i$$

Now the excess of the rest mass is transformed into the kinetic energy of the subsystems. This is examplified by the fission of atomic nuclei. In this case, a nucleus with a certain given nucleon number transforms itself into a state with less energy (hence a more stable state), while decaying into nuclei with masses M_i that have a lower nucleon number. The kinetic energy liberated in this process is balanced by the mass surplus $M - \sum M_i > 0$.

One can also describe the condition for instability by referring to the mass defects of the participating nuclei. While it is true that a nucleus with a positive mass defect $\Delta M = \sum_i m_i - M > 0$ is stable against decay into individual protons and neutrons (masses m_i), this does not imply that it is stable in an absolute sense. It could decay into lighter nuclei with masses M_i, where these nuclei have certain appropriate binding energies, i.e. mass defects ΔM_i. Clearly, the system after the decay has a more stable configuration than the original nucleus if the mass defect ΔM of the nucleus in question is smaller than the sum of the mass defects of the lighter nuclei, i.e. if

$$\Delta M < \sum_i \Delta M_i \tag{5.19}$$

Since to a smaller mass defect there corresponds a larger rest mass, the two conditions of instability are equivalent. Thus, the mass defects permit the prediction of stability or instability of isotopes. For example, the nucleus ${}^{9}_{4}\text{Be}$ is absolutely stable. Not only is its mass defect positive, but it is also larger than the sum of mass defects of all nuclei into which it could decay. In contrast, the nucleus ${}^{8}_{4}\text{Be}$, while still stable against decay into individual nucleons, may spontaneously decay into two α-particles (${}^{4}_{2}\text{He}$).

In any case, the mass equivalent of energies measured in the usual units is very small. Therefore it is not surprising that the mass-energy relation was discovered so late. Indeed, to give an example, 2500 kWh is equivalent to only the tiny mass of 10^{-4} g, provided this could be totally converted into energy.[1] In fact, this occurs only in the case of a particle-antiparticle annihilation or in the case of the decay of electrically neutral particles into photons (such as $\pi^{o} \rightarrow 2\gamma$). In all other cases only a fraction of the rest mass is liberated as energy.

In this respect it is interesting to observe that the human body's sensitivity for energy (such as to electromagnetic radiation perceived by sight) is many orders of magnitude higher than its sensitivity for registering inertial effects of masses. In fact, the maximum sensitivity of the retina occurs at a wavelength of about 5×10^{-5} cm and for the threshold excitation of the sensory cells one needs about 5 light quanta of this wavelength. Thus, visibility threshold (scotopic vision) starts at 10^{-11} erg of incident radiation energy. On the other hand, the sensory perception of 1g requires the energy amount of 10^{21} erg. Accordingly the human sensory organs are 10^{32} times more sensitive to the effects of radiative energy than to the inertial or gravitational effect of masses. But suppose that, instead, the sensitivities would be of the same order of magnitude. Then one would feel also the inertial effects of the photons that hit the body and the equivalence of mass and energy would be a selfevident matter of experience.[2]

Under the prevailing realistic conditions it was of course necessary to develop methods to determine minute mass differences, so as to eventually test the equivalence of mass and energy .in reactions where a sufficient amount of rest mass is transformed into energy.

---

[1]
These figures follow from $1g(3 \times 10^{10})^2$ $(cm\ sec^{-1})^2 = 9 \times 10^{20}$ erg $= 9 \times 10^{13}$ J, whereby the value of the speed of light conveniently has been rounded off.

[2]
This was pointed out by F. Dessauer in Helv. Phys. Acta <u>15</u>, 108 (1942).

In exothermal chemical reactions, which of course are connected with processes in the atomic envelopes, only extremely small amounts of mass are liberated as energy. For example, the mass defect related to the burning of 1 kg carbon is 4×10^{-7} g, so that the relative mass defect $\Delta M/M$ becomes only 4×10^{-10}. The ultimate reason is that, because of the smallness of the fine structure constant and the electron-proton mass ratio, atomic binding energies are very small.

In contrast, mass defects corresponding to the binding energies of atomic nuclei have measurable magnitudes. In these cases, the relative mass defect is almost 1%. The liberation of the equivalent energy is the basis of energy production in nuclear fission and fusion. The enormous energy radiation of stars (such as the Sun) is made possible by a number of nuclear reactions which proceed in the stellar interiors at the approximate temperature of 10^7 K. Particularly important is the chain of reactions in which eventually helium nuclei (^{4}He) are produced by way of fusion of hydrogen nuclei (^{1}H). In this manner, a rich reserve of "nuclear fuel" secures the expected long lifetime of the Sun. Assuming a radiative loss of 4.4×10^{12} g per second, the Sun will lose by radiation only 0.07% of its mass in 10^{10} years, provided its light emission power remains constant.

Even higher mass defects arise in consequence of the gravitational binding energy when one considers star formation. For example, in the case of neutron stars one may have as much as a 30% relative mass defect. Finally, in case of matter-antimatter annihilation, one has a change of total rest energy ($2mc^2$) into the equivalent radiative energy.

One should not misunderstand the often used description of these processes in terms of "transmutation of matter into energy". For example, in nuclear fusion the number of nucleons remains unchanged, and in the transition into the energetically more advantageous state only a fraction of the total mass is liberated as energy. Besides, the photons of the emitted electromagnetic radiation are also elementary particles. Like all matter, they have inertia. Thus, basically it is only the form of matter that changes; due to the universal equivalence, it may manifest itself either as mass or as energy.

In summary, the fundamental relevance of Einstein's mass-energy relation is the possibility of changing rest mass into its equivalent amount of energy and vice versa. But this does not mean that these two concepts can be equated. They differ in their behaviour under Lorentz transformations. The rest mass stands for the amount of the four-momentum, whereas the energy is a measure of its temporal component. Correspondingly, the energy of a closed system is conserved. In contrast, there is no conservation law for the rest mass.

5.3   Interactions of Relativistic Particles via Fields

Formulating the dynamics of a system consisting of a number of relativistic particles is a rather complex problem. It is best to account for the finite propagation velocity of all effects by introducing a field that mediates the interaction, which, therefore assumes physical reality. This procedure goes beyond the framework of particle mechanics. It is true that, besides the interaction mediated by fields, in principle one also can construct relativistic action-at-a-distance theories (where retardation holds). However, these theories did not compete successfully with the theories that are based on interactions mediated by fields ("close-range action").

According to equation (5.4) the motion of a sytem of particles is determined by the Minkowski forces K^{μ}_{i} which act on the individual particles:

$$\frac{d}{d\tau_i} \, p^{\mu}_{\ i} = K^{\mu}_{\ i} \, , \quad i = 1, 2, \ldots, N \ .$$

Here the differential relation between the proper times τ_i of the N particles and the time t in the chosen coordinate system K is given by

$$d\tau_i = \sqrt{1 - \frac{v_i^2}{c^2}} \ dt$$

These, so far formal equations, determine the dynamical problem only if the acting forces K^{μ}_{i} are known. In case of a closed system (no external forces) one has only internal forces between the particles. Obviously they originate from the particles themselves.

We know four fundamental interactions of varying strengths: these are the gravitational, weak, electromagnetic, and strong interactions. The gravitational and electromagnetic interactions have a long range and hence they can be experienced macroscopically (i.e., classically). However, the other two interactions act only at distances of about 10^{-13}cm or less. But at such small dimensions the classical concept of an orbit becomes meaningless and it is no longer possible to describe the processes with classical notions such as force and acceleration. In the microscopic domain one must use quantum theory. This gives rise to experimentally verifiable predictions of probabilities for scattering, particle transmutations, and properties of bound states.

For the high particle velocities that occur in relativistic physics, one must take into consideration the finite propagation velocity of effects. Unlike in Newtonian mechanics, instantaneous interaction is not possible. One must take into consideration that an effect experienced at the world point x_2 must have originated at the world point x_1 at a time $|\vec{x}_2 - \vec{x}_1|/c$ earlier, corresponding to its propagation time. (This is called retardation.) Therefore, the interaction will depend in a complicated way on the motions of the relevant particles. For this reason it is best to describe interactions of relativistic particles in terms of fields which propagate in space.

In Newtonian mechanics the instantaneous force field, depending only on distances, may be taken to be a convenient means for describing the dynamics. But in relativity theory, the change in the position of one particle influences another particle only after the perturbation of the field in its neighborhood propagated through the distance which separates the particles. Thus, the field produced by the particle assumes a physical reality of its own. This is examplified by the electromagnetic field. The particles in motion are the sources of the field, and in course of the interaction, energy and momentum is transferred between the particles by the vehicle of the field. The changes of energy-momentum propagate in this field with the maximal speed of action. Thus, the field which transmits the interaction carries itself energy and momentum. In that way its physical reality finds expression.

If only internal forces act, the total momentum of the system does not change. But since the closed system consists both of the particles and the field which transmits the interaction, in the balance of momentum one must take into account the momentum of the field, i.e.

$$\sum_i p_i^\mu + P_{Feld}^\mu = \text{const} \tag{5.20}$$

Thus, the total momentum of the interacting particles as such is not conserved. From this one concludes that the mutual forces between the particles (action and reaction) are not "equal and opposite".[1] The third axiom of Newtonian mechanics (actio est reactio) does not hold in general. Only if the duration of the interaction is negligible, for instance in the impact approximation, is one permitted to neglect the contribution of the field momentum and have momentum conservation satisfied by considering the four-momenta of the particles alone. We shall return to this point when we discuss reactions between particles.

Among the four fundamental interactions only those which have a long range can be described classically. Since gravitational fields lead to a position dependent metric $g_{\mu\nu}(x)$ and thus alter the space-time structure, their discussion goes beyond the special theory of relativity. They are studied in the framework of general relativity. The classical electromagnetic field will be discussed in Chapter 6.

We wish to illustrate the role of the field concept which extends beyond the realm of classical physics in the following comments. In the classical theory particles are described by their coordinates $x^\mu(\tau)$ and fields by the corresponding functions $A(x)$, that is, in a completely different manner. In the microscopic domain, i.e.

---

[1] One should recall the proof of momentum conservation for a system where only internal forces are present: here, in Newtonian mechanics, the third axiom is used.

in quantum theory, this asymmetric treatment is eliminated. Since quantization of a field leads to discret quantum states, one may associate with the fields corresponding particles. Thus, in quantum theory of fields one has a synthesis of particles and fields. Here both are described by field operators $\psi(x)$, $A(x)$, that is to say, one deals only with fields and with the particles corresponding to the fields, respectively.

By introducing the field concept one went beyond the frame of particle mechanics proper. It is of fundamental interest to see if it were possible to formulate the dynamics of a system of inter-acting particles which does not use fields yet is consistent with special relativity. For the electromagnetic interaction between charged particles Wheeler and Feynman were able to formulate a relativistically invariant action integral in which the electromagnetic field does not occur but which nevertheless describes the phenomena of classical electrodynamics.[1] This proposal definitely has a historical signifi-cance. But nowadays the motivations for such a relativistic action-at-a-distance theory are less convincing. In particular, in view of the independent reality of the radiation field, proved by the existence of the photon, the complete elimination of the field appears arbitrary and hardly justified. None the less one ought not simply overlook the possiblity of formulating electromagnetic interactions in such a manner.[2]

However, one runs into difficulties if one tries to construct a Hamiltonian formulation of relativistic action-at-a-distance, even though this would be useful to perform quantization later on. Indeed, if the coordinates and momenta of particles were canonical variables and transformed correctly under a Lorentz transformation

---

[1] J.A. Wheeler and R.P. Feynman, Rev. Mod. Phys. 17, 157(1945); 21, 424(1949).

[2] Lately, the notion of a direct-interaction theory was again dis-cussed by F. Hoyle and J.V. Narlikar: Action at a Distance in Physics and Cosmology (W.H. Freeman, San Francisco, 1974).

then there could not be any interaction between the particles.[1] Several attempts were made to see if it were possible to circumvent this "no-interaction-theorem" by relaxing the assumptions.[2] One interesting proposal for solving the problem was put forward by F. Rohrlich, in which the internal dynamics of a system of relativistic particles is described in terms of a Hamiltonian theory with constraints.[3]

Substantially easier is the formulation of relativistic dynamics for a system of charged particles with moderate velocities, where one may neglect retardation effects of the electromagnetic field. For such a system one can construct a Lagrangian which depends only on the instantaneous positions and velocities of the particles. This approximative Lagrangian describes the interacting particles correctly up to and including contributions of order $(v/c)^2$.[4]

Particularly simple is the case when most of the time the particles are far away from each other and are moving with high velocities. For in that case the individual particle energies

$$E_i = \frac{m_i c^2}{\sqrt{1 - \beta_i^2}}$$

are large compared to the interaction energies at great distances. Therefore, the latter can be neglected and so the system may be treated

[1] D.G. Currie, T.F. Jordan, and E.C.G. Sudarshan, Rev. Mod.Phys. 35 , 350 (1963); H. Leutwyler, Nuovo Cim. 37, 556 (1965).

[2] See, for example, R.A. Mann: The Classical Dynamics of Particles (Academic Press, New York, 1974), Chapter 5, and the literature quoted therein. A selection of studies relevant in this area is reprinted in E.H. Kerner: The Theory of Action at a Distance in Relativistic Particle Dynamics (Gordon and Breach, New York, 1972).

[3] F. Rohrlich, Phys. Letters 66A, 268 (1978); Annals of Physics 117, 292 (1979). Reviews and further references are contained in: Relativistic Action at a Distance: Classical and Quantum Aspects, J. Llosa, Ed. (Springer, Berlin, 1982).

[4] A discussion of this Darwin Lagrangian can be found, for example, in J.D. Jackson: Classical Electrodynamics, 2nd ed. (J. Wiley, New York, 1975), p. 593.

as one consisting of free particles. This kind of situation pertains clearly to the collision of relativistic particles. Before the collision the particles are far from each other and therefore they are "free". In course of the collision they interact for a short timespan (the details of the interaction are here of no interest), and this takes place in a small region of space. Then they fly apart (in reactions these are, generally speaking, particles different from those before collision), and, when they are sufficiently distant from one another, they again move as free particles. Because of momentum conservation the total momentum of the free moving particles before collision must be equal to the total momentum of the free particles after the collision. Therefore, in this case one can write down the energy—momentum conservation law without a contribution from the fields in the following form:

$$\sum_{i=1}^{N} p_i^{\mu} = \sum_{k=1}^{N'} p_k'^{\mu} \tag{5.21}$$

Because of the possibility of production or annihilation of particles, the particle numbers N before and N' after the reaction has taken place, may be different.

Equation (5.21) has abundant applications for reactions between particles. This law has been verified experimentally over and over again. It has a fundamental significance for nuclear and elementary particle physics. In the next Chapter we discuss, out of the many processes where relativistic effects play an important role, some particularly typical cases.

5.4   Energy-Momentum Conservation for Particle Processes

For the study of particle reactions it is often advantageous to use a reference frame in which the spatial part of the total momentum is zero, i.e. $\vec{P} = 0$. On account of the timelike character of momentum, it is always possible to find such a system. As in classical mechanics, such a system is called the center-of-mass system. The experimental situation however, corresponds to a frame of reference in which the experimenter and his instruments are at rest – this is called the laboratory system. The transition from one system to the other is rendered by a Lorentz transformation.

If one forms in the center-of-mass system K^* the square of the total momentum,

$$\frac{E^{*2}}{c^2} - \vec{P}^{*2} = M^2 c^2$$

then, on account of $\vec{P}^* = 0$ it follows that

$$E^* = M c^2$$

Here M is the rest mass of the system. On the other hand, the energy E^* is composed from the energies of the individual particles with velocities v_i relative to the center of mass, according to the formula

$$E^* = \sum_i \frac{m_i c^2}{\sqrt{1 - \frac{v_i^2}{c^2}}}$$

Therefore we get for the rest mass

$$M = \sum_i \frac{m_i}{\sqrt{1 - \frac{v_i^2}{c^2}}}$$

From this we immediately see that the rest mass of the particle system is not simply equal to the sum of the rest masses of the individual particles but it also depends on their velocities in the center-of-mass system. If, for example, one envisages a relativistic ideal gas, then the rest mass M of this system will depend on the internal motion of the gas particles, or, equivalently, on the gas temperature. More generally, the rest energy of a composite system contains the entire internal energy (cf. page 115).

5.4.1 Decay

Suppose the spontaneous decay of a particle (or of a system) with mass M gives rise to two particles with masses m_1, m_2 and momenta $\vec{p}_1$, $\vec{p}_2$, respectively. In the center-of-mass system, which in this case coincides with the rest system K of the decaying particle, the spatial parts of the total momentum before the decay (P) and after the decay (P') are both zero:

$$\vec{P} = 0 \; , \quad \vec{P'} = \vec{p}_1 + \vec{p}_2 = 0$$

Because of momentum conservation, $P = P' = p_1 + p_2$, the invariant squares of the four-vectors P and P' must also be equated:

$$M^2 c^2 = \left(\frac{E_1 + E_2}{c} \right)^2 - \left(\vec{p}_1 + \vec{p}_2 \right)^2$$

Using $\vec{p}_1 + \vec{p}_2 = 0$, in the center-of-mass system this yields

$$M c^2 = E_1 + E_2$$

or, separating out the rest energies,

$$M c^2 = m_1 c^2 + m_2 c^2 + E_{1\,kin} + E_{2\,kin}$$

Since the kinetic energies of the emitted particles are positive, it then follows that the spontaneous decay of a particle with mass M is energetically possible only if the inequality

$$M > m_1 + m_2$$

is satisfied. Conversely, if the mass of the particle is less than the sum of the rest masses of the decay products, then spontaneous decay is not possible. In this situation decay could occur only if sufficient energy were supplied from outside.

Next, let us determine the energies of the decay products. According to the relation

$$E = \sqrt{\vec{p}^2 c^2 + m^2 c^4}$$

one obtains for E_2, if one uses $\vec{p}_1^{\,2} = \vec{p}_2^{\,2}$, the value

$$E_2^2 = \vec{p}_2^2 c^2 + m_2^2 c^4 = E_1^2 - m_1^2 c^4 + m_2^2 c^4$$

Because of

$$E_1^2 = \left(M c^2 - E_2 \right)^2$$

we can eliminate $E_1{}^2$, so that E_2 becomes expressed in terms of the masses of the participating particles:

$$E_2 = \frac{(M^2 - m_1{}^2 + m_2{}^2)\, c^2}{2M}$$

Similarly, with an interchange of indices,

$$E_1 = \frac{(M^2 - m_2{}^2 + m_1{}^2)\, c^2}{2M}$$

Substracting the rest energy, we finally get

$$E_{1\,kin} = \frac{(M - m_1)^2\, c^2 - m_2{}^2 c^2}{2M} \tag{5.22}$$

Thus, for a decay into two particles, the center-of-mass system kinetic energies of the decay products are determined entirely by the masses of the participating particles and therefore they have fixed values.

As an example, let us consider the already mentioned decay of the charged π-meson into a muon ($\mu^{\pm}$) and a corresponding neutrino (ν_μ) (see p.41). According to equation (5.22), we get for the kinetic energy of the muon

$$E_{\mu\,kin} = \frac{(m_\pi - m_\mu)^2\, c^2 - m_\nu{}^2 c^2}{2 m_\pi}$$

where the different masses of the particles are denoted by indices. Now, the kinetic energy of the charged muon can be determined from ionization it causes in the detecting device. From this measured value and from the known masses m_π and m_ν it follows that the neutrino mass must be very small compared to that of the pion. In general, one takes $m_\nu = 0$. But because of the limited experimental accuracy the discussed decay allows only for ascertaining an upper limit of the muonic neutrino mass. Presently, this value is $m_\nu < 0.25$ MeV.[1]

---

[1] See Particle Data Group, Phys. Lett. 204B (1988), p. 13, and ibid, p. 139 for further references. One actually supposes that the neutrino has a mass that lies well below this upper limit. The comparison with the upper limit for the photon mass (p.29) is instructive. The latter is about 3×10^{-27} eV, i.e. 32 orders of magnitude less.

On the other hand, for neutron decay, with a mean life of about 15 minutes, the kinetic energy of the emitted electrons is not fixed like this; one rather finds an energy distribution. Hence it follows that here we cannot have a decay into only two particles (electron and proton). But a third charged particle is forbidden by the conservation of electric charge, so that the third particle, which takes away part of the energy, must be electrically neutral. These facts led W. Pauli to predicting the existence of the neutrino (1930). Only in 1956 was the experimental proof of the existence of the neutrino actually provided; this was made possible by the high neutrino intensities that occur in nuclear reactors.

5.4.2 Creation

Besides nuclear fission and fusion, the most significant consequence of the energy-mass equivalence is the possibility of creating new particles. Such creation processes are studied in high energy laboratories with the huge particle accelerators. For the creation of a particle with rest mass m, one obviously needs at least an energy of mc^2. Actually, more energy is required, because in many cases one must also satisfy other conservation laws which concern certain charge-like quantum numbers. For this reason, particles can be created only in pairs, with opposite charges. For example, to create an electron-positron pair ($\gamma \rightarrow e^- + e^+$) one should need at least the energy $2m_ec^2$. As we shall see shortly, the threshold energy for this process is in fact still somewhat higher.

Apart from the reason just given, a higher energy for creation processes is often necessitated by the experimental set-up. For, if the target is at rest in the laboratory, it will take up energy in course of the collision and then moves on with a certain velocity. Consequently, the energy amount required for this cannot be used for the production of new particles. However, if one arranges that two particle beams with equal and opposite momenta collide, then (because of $\vec{p}_1 + \vec{p}_2 = 0$) in the center-of-mass system of the collision the entire kinetic energy of the initial state is available for the production of particles. This more effective utilization of the invested energy is a special advantage of storage rings.

By way of example we consider electron-positron pair creation by photons. To start with, one sees that this process cannot occur in vacuum. Before the reaction the total momentum is lightlike ($q^2 = 0$), but afterwards it is timelike because the sum of two timelike vectors with equal sign of the zero-component is again timelike. Consequently one has $q^2 \neq (p_1 + p_2)^2$. But this result contradicts energy-momentum conservation. Hence pair creation can occur only in the presence of an additional particle, which usually is an atomic nucleus.

Now we determine the threshold energy $\tilde{E}$ for this process. We assume that photons with energy $\tilde{E}$, hence, according to equation (5.13), with the corresponding momentum $|\vec{q}| = \tilde{E}/c$ impinge on a nucleus of mass M at rest. Thus, before the reaction the squared total momentum is

$$P^2 = \left(\frac{\tilde{E} + Mc^2}{c} \right)^2 - \left(\frac{\tilde{E}}{c} \right)^2$$

Because of momentum conservation (P=P'), this is the same as the squared total momentum P'^2 after the reaction. The value of the invariant $c^2 P'^2$ may be calculated in any arbitrary inertial system. Since in the center-of-mass system of the particles $\vec{P}' = 0$, we obtain

$$c^2 P'^2 = (E_{Kern} + E_{e^+} + E_{e^-})^2$$

The threshold energy $\tilde{E}$ corresponds clearly to the situation where the particles in the final state have no kinetic energy. Taking this into account in the expression of $c^2 P'^2$ and remembering that $P^2 = P'^2$, it follows that

$$(\tilde{E} + Mc^2)^2 - \tilde{E}^2 = (Mc^2 + 2m_e c^2)^2$$

hence the threshold energy is

$$\tilde{E} = 2 m_e c^2 \left(1 + \frac{m_e}{M} \right) \tag{5.23}$$

We see that this is greater than $2m_e c^2$. In case of $m_e/M \ll 1$, the second term in the parenthesis may be neglected. For pair creation in the field of a nucleus this is a good approximation. But if the process occurs in the field of an electron ($M=m_e$) then one has the threshold energy $\tilde{E}=4m_e c^2$.

The process inverse to pair creation is pair annihilation,

$$e^+ + e^- \longrightarrow 2\gamma$$

Because of momentum conservation there must be (at least) two photons emerging. Let now p_1 and p_2 be the momenta of the elctron and of the positron, respectively, and let q_1 and q_2 be the momenta of the two photons. In the center-of-mass system of the initial state one has $\vec{p}_1 = -\vec{p}_2$. On account of momentum conservation, this leads to $\vec{q}_1 = -\vec{q}_2$, which means that the two photons are emitted into opposite directions with equal momenta, hence also with equal energies. From energy conservation it follows that the energy of each photon is

$$E_\gamma = c\sqrt{\vec{p}_1^{\,2} + m_e^2 c^2}$$

The smallest possible photon energy is obtained if $\vec{p}_1 = 0$; it equals $m_e c^2$. In pair annihilation the rest energy of the particles is completely transformed into the energy of the produced photons. This process is a particularly impressive example of the mass-energy relation. Nevertheless, matter does not "disappear" here, since the originating photons are certainly to be considered material objects. At any rate, they are distinguished by the fact that they have zero rest mass.

5.4.3 Scattering

If in a collision of two particles a and b these particles run out without transmutation or some inner excitation and simply change their momenta, the process is called elastic scattering. The kinetic energy available in the initial state does not appear after the collision in the form of rest energy or excitation energy, but rather it is again distributed as kinetic energy among the particles of the final state. Referring to the balance of four-momenta before and after scattering, i.e. due to

$$p_a + p_b = p_a' + p_b'$$

one can write down invariant relations between expressions of the form $p_a \cdot p_b'$, where $p_a \cdot p_b'$ stands for the inner product of the vectors p_a and p_b'. In this manner one can determine in the laboratory or center-of-mass system interesting quantities such as the scattering angle or the particle energies and their relations to each other.[1]

As an interesting example we consider the Compton effect (A.H. Compton, 1922), which, by its simple interpretation as an elastic collision between a light quantum (photon) and an electron verifies the quantal nature of electromagnetic radiation in an impressive way. Since the electron experiences a recoil (i.e., takes up energy), the scattering photon has less energy than the incident photon, and therefore, in accord with the relation $E_\gamma = h\nu = hc/\lambda$ (where h is the Planck constant), it has a longer wavelength.

Since photons propagate with the speed of light, the Compton effect is a relativistic collision process. Let the four-momenta of the photon and of the electron be, before and after the scattering, q, q' and p, p', respectively. Momentum conservation requires

$$q + p = q' + p'$$

For the calculation of the photon's energy change, one expresses from this equation the momentum p' of the recoiling electron and takes the square:

$$(q + p - q')^2 = p'^2$$

Since $p^2 = p'^2 = m_e^2 c^2$ and $q^2 = q'^2 = 0$, it then follows that

$$q \cdot q' = p \cdot (q - q')$$

We now go to the rest system of the electron, where $p = (m_e c, 0, 0, 0)$. Let us denote the directions of the incident and scattered photon by $\vec{n}$ and $\vec{n}'$, respectively. The angle between these directions (i.e., the scattering angle of the photon) is defined by $\vec{n} \cdot \vec{n}' = \cos \Theta$.

---

[1] Further details of such calculations will not be given here. See in this connection, e.g. the book by E. Byckling and K. Kajantie, Particle Kinematics (J. Wiley, New York, 1973).

With the values $q=(1,\vec{n})E_\gamma/c$ and $q'=(1,\vec{n}')E'_\gamma/c$ of the photon momenta then follows that

$$\frac{1}{E'_\gamma} - \frac{1}{E_\gamma} = \frac{1}{m_e c^2}(1-\cos\theta)$$

But in consequence of the formula $E_\gamma=h\nu=hc/\lambda$, to a light quantum energy E_γ there corresponds the wavelength λ. Therefore, the change of wavelength in Compton scattering is

$$\Delta\lambda = \lambda' - \lambda = \frac{h}{m_e c}(1-\cos\theta) \qquad (5.24)$$

In accord with experiment this result shows that $\Delta\lambda$ depends only on the scattering angle and on the mass of the scattering particle but not on the wavelength of the incident light. The maximal possible change of wavelength is 2Λ where $\Lambda= h/m_e c=2.4\times10^{-10}$ cm is called the Compton wavelength of the electron. It amounts to the wavelength of a light quantum with an energy equal to the rest energy of the electron. By measuring the Compton wavelength one can determine the Planck constant h. In order to make the effect well pronounced, the wavelength of the incident radiation should be comparable with the Compton wavelength of the electron. Therefore, to demonstrate the effect, short-wave X-rays or γ-rays are particularly suitable.

From equation (5.24) it follows that, in the rest system of the electron and if $\theta\neq0$, one always has an increase of the wavelength. However, in a reference system which is in motion relative to the electron (i.e., when light is scattered off a moving electron), one also may have a decrease of wavelength. In that case, energy is transferred from the electron to the photon. The inverse Compton effect has astrophysical significance.

5.5 The Principle of Least Action; the Lagrangian

In this Section we return to the relativistic mechanics of a free particle and will reformulate it in a covariant language by introducing an appropriate Lagrangian. This discussion serves also as a preparation for the following chapter, in which covariant electrodynamics is to be treated.

According to Hamilton's principle each mechanical system possesses an action integral S which becomes extremal for the actual motion, so that its variation S corresponding to the transition to another infinitesimally near path vanishes. We will apply this principle for a free particle.

Because of the universal relativity principle the action integral must not depend on the choice of the inertial system. Therefore, it must be invariant with respect to Lorentz transformations. When we discussed on page 103 the concept of proper time, we already saw that the proper time, or equivalently the separation of the two world points, possesses a maximum along a straight world line (i.e. for a free particle). This observation motivates the following Ansatz for the action integral of a free particle:

$$S = -\alpha \int_1^2 ds \qquad (5.25)$$

where α is a positive constant yet to be determined. This constant will characterize the particle in question. The negative sign has the effect that the maximum along a straight world line becomes a minimum, and that for small velocities S goes over into the known Newtonian expression with the correct sign.

Let the world line $x^\mu(\lambda)$ of the particle depend on an invariant parameter λ which, as a function of the proper time τ, increases monotonically with τ, but otherwise is arbitrary. Then the line element becomes

$$ds = \sqrt{g_{\mu\nu} \frac{dx^\mu}{d\lambda} \frac{dx^\nu}{d\lambda}}\, d\lambda \qquad (5.26)$$

The quantities to be varied are the coordinates x^μ and the derivatives

$$\frac{dx^\mu}{d\lambda} = \dot{x}^\mu$$

which for $\lambda = \tau$ become the four-velocity

$$\frac{dx^\mu}{d\tau} = u^\mu$$

For the time being we take λ different from τ. Otherwise only such variations would be permitted which are compatible with the condition

$$g_{\mu\nu}\frac{dx^\mu}{d\tau}\frac{dx^\nu}{d\tau} = u_\mu u^\mu = c^2$$

and so one ought to do the variational calculation with this subsidiary condition (i.e. with a Lagrange multiplier). This is avoided by identifying λ with the proper time τ only after the variation has been performed, and then sets

$$\sqrt{g_{\mu\nu}\,\dot{x}^\mu\dot{x}^\nu}\,d\lambda = c\,d\tau \tag{5.27}$$

From the condition of an extremum, $\delta S = 0$, one obtains for variations which vanish at the end-points λ_1 and λ_2 the Euler-Lagrange equations of motion:

$$\frac{d}{d\lambda}\frac{\partial L}{\partial\dot{x}^\mu} - \frac{\partial L}{\partial x^\mu} = 0 \tag{5.28}$$

The integrand of the action integral S, the Lagrangian

$$L = -\alpha\sqrt{g_{\mu\nu}\dot{x}^\mu\dot{x}^\nu} \tag{5.29}$$

does not depend on x^μ so that (5.28) gives

$$\alpha\frac{d}{d\lambda}\left[\frac{\dot{x}^\mu}{\sqrt{g_{\mu\nu}\dot{x}^\mu\dot{x}^\nu}}\right] = 0$$

Now we identify the parameter λ of the world line for the actual motion with the proper time τ and obtain, using (5.27), the equation of motion for a free particle

$$\frac{d^2x^\mu}{d\tau^2} = 0$$

The constant factor α omitted here is readily determined to be $\alpha = mc$. On account of the relation (4.6) the action S may be represented

136

as an integral over the time t :

$$S = -\alpha c \int_{t_1}^{t_2} \sqrt{1 - \frac{v^2}{c^2}} \, dt$$

Thus, the Lagrangian expressed in terms of the ordinary velocity v is [1]

$$\tilde{L} = -\alpha c \sqrt{1 - \frac{v^2}{c^2}}$$

For small velocities $v \ll c$ it follows that

$$\tilde{L} \approx -\alpha c + \alpha c \frac{1}{2} \frac{v^2}{c^2}$$

Apart from the constant term $-\alpha c$ which is irrelevant for the equations of motion, this expression coincides with the nonrelativistic Lagrangian $mv^2/2$, provided one sets $\alpha = mc$. Thus, the constant α is determined by the rest mass which characterizes the particle. In conclusion, the Lagrangian of a free particle can be written as

$$\tilde{L} = -mc^2 \sqrt{1 - \frac{v^2}{c^2}} \qquad (5.30)$$

From this one obtains the spatial momentum components

$$\vec{p} = \frac{\partial \tilde{L}}{\partial \vec{v}} = \frac{m\vec{v}}{\sqrt{1 - \frac{v^2}{c^2}}}$$

Using the connection of the Hamiltonian with the Lagrangian, one obtains from (5.30) the energy,

$$E = \vec{p} \cdot \vec{v} - \tilde{L}$$

$$= \frac{mc^2}{\sqrt{1 - \frac{v^2}{c^2}}}$$

---

[1] It should be noted here, that the Langrangian $\tilde{L}$ is different from L(equ. (5.29)). In fact, while the latter as a Lorentz invariant refers to the integration parameter λ (or τ), the former, associated with t as integration variable, is not an invariant quantity.

In this manner, the earlier introduced quantities which were discussed in Section 5.2, can be derived from the invariant action integral. According to equation (5.12), the energy of a free particle can be expressed in terms of its momentum and rest mass. This then yields the Hamiltonian:

$$H = c \sqrt{\vec{p}^2 + m^2 c^2}$$

For momenta $\vec{p}^2 \ll m^2 c^2$, i.e. for small velocities it follows that

$$H \approx m c^2 + \frac{\vec{p}^2}{2m}$$

Up to the rest energy, this is the Newtonian expression for the Hamiltonian of a free particle.

For particles with zero rest mass the Hamiltonian assumes the particularly simple form

$$H = |\vec{p}| c$$

This relation is approximately valid also for other particles whenever $\vec{p}^2 \gg m^2 c^2$.

5.6   Conservation Laws

The conservation laws of four-momentum and of the four-dimensional angular momentum tensor follow from the invariance of a closed system against translations and against homogeneous Lorentz transformations, respectively. Under these symmetry transformations the action integral stays unchanged, because a dynamically possible orbit $x^\mu(\lambda)$ should go over into another one $x'^\mu(\lambda)$ in the "dashed" reference system K'. Thus the corresponding infinitesimal transformation $x'^\mu = x^\mu + \delta x^\mu$ leads to $\delta S = S' - S = 0$. Performing the symmetry transformation, this then gives

$$\delta S = \int_{\lambda_1}^{\lambda_2} \left\{ \frac{\partial L}{\partial x^\mu} \, \delta x^\mu + \frac{\partial L}{\partial \dot{x}^\mu} \, \delta(\dot{x}^\mu) \right\} d\lambda = 0$$

After integration by parts of the second term and using

$$\delta(\dot{x}^{\mu}) = \frac{d}{d\lambda}(\delta x^{\mu})$$

it follows that

$$\delta S = \frac{\partial L}{\partial(\dot{x}^{\mu})} \delta x^{\mu}\bigg|_{\lambda_1}^{\lambda_2} - \int_{\lambda_1}^{\lambda_2} \left\{ \frac{d}{d\lambda} \frac{\partial L}{\partial(\dot{x}^{\mu})} - \frac{\partial L}{\partial x^{\mu}} \right\} \delta x^{\mu} \, d\lambda = 0$$

Since with the above symmetry transformation $x'^{\mu} = x^{\mu} + \delta x^{\mu}$ we look at the actual path followed by the system, the equations of motion (5.28) are satisfied, and the change of the action inegral is given by [1]

$$\delta S = \frac{\partial L}{\partial(\dot{x}^{\mu})} \delta x^{\mu}\bigg|_{\lambda_1}^{\lambda_2} = 0$$

After the calculation is performed, one may set again $\lambda = \tau$ along the actual world line and thus one gets

$$\delta S = \frac{\partial L}{\partial u^{\mu}} \delta x^{\mu}\bigg|_{\tau_1}^{\tau_2} = 0$$

With the definition

$$p_{\mu} = -\frac{\partial L}{\partial u^{\mu}} \tag{5.31}$$

of the generalized (i.e. canonical conjugate) momentum this yields

$$\delta S = -p_{\mu} \delta x^{\mu}\bigg|_{\tau_1}^{\tau_2} = 0 \tag{5.32}$$

If one sets the δx^{μ} equal to the infinitesimal constant parameters

---

[1] One must observe that δx^{μ} representing here an infinitesimal Lorentz transformation into the reference system K' does not vanish at the end-points λ_1 and λ_2 of the world line. Despite the identical notation, these changes of the coordinates should therefore not be confused with the variations which were used in connection with the extremum principle. In the letter case, these variations served to compare different orbits which connect fixed end-points.

of a translation $\delta\epsilon^\mu$, then it follows that

$$p_\mu \, \delta\epsilon^\mu \, \Big|_{\tau_1}^{\tau_2} = 0 \qquad (5.33)$$

This shows that along the world line taken by the particle, its momentum p_μ is conserved. For a free particle, using (5.29), one obtains the earlier expression (see p.106) of the momentum components:

$$p^\mu = m \, u^\mu$$

In an analogous manner the invariance against homogeneous Lorentz transformations leads to the conservation of the angular momentum tensor. So as to assure that, under Lorentz transformations $\delta x^\mu = \delta\epsilon^\mu_{\ \nu} x^\mu$ the square of the four-vector x^μ stays invariant, i.e. that

$$(x^\mu + \delta\epsilon^\mu_{\ \nu} x^\nu)(x_\mu + \delta\epsilon_\mu^{\ \nu} x_\nu) = x^\mu x_\mu$$

the infinitesimal parameters $\delta\epsilon^\mu_{\ \nu}$ must form an antisymmetric tensor, $\delta\epsilon^\mu_{\ \nu} = -\delta\epsilon_\nu^{\ \mu}$. To start with, according to (5.32) one gets

$$p_\mu x^\nu \, \delta\epsilon^\mu_{\ \nu} \, \Big|_{\tau_1}^{\tau_2} = 0$$

The tensor $p_\mu x^\nu$ can be decomposed into a symmetric and into an anti-symmetric part (see p.67). Since the contraction of the symmetric part with the antisymmetric tensor $\delta\epsilon^\mu_{\ \nu}$ vanishes, it follows that

$$\frac{1}{2}\left(p_\mu x^\nu - p^\nu x_\mu \right)\delta\epsilon^\mu_{\ \nu} \, \Big|_{\tau_1}^{\tau_2} = 0$$

Hence, the angular momentum tensor

$$L^{\mu\nu} = x^\mu p^\nu - x^\nu p^\mu \qquad (5.34)$$

is conserved along the world line followed by a closed system. Its spatial components L^{mn} yield precisely the three-dimensional angular momentum tensor $\vec{l} = \vec{x} \times \vec{p}$,

$$L^{12} = x^1 p^2 - x^2 p^1 = \ell^3, \text{ etc.}$$

For the mixed components L^{01}, L^{02}, L^{03} the above conservation law leads to the constant vector

$$c \left(t \vec{p} - \frac{E}{c^2} \vec{x} \right) = \text{const}$$

Since E and $\vec{p}$ are conserved, we see that the closed system, considered as a whole, moves with the constant velocity

$$\frac{d\vec{x}}{dt} = \frac{c^2 \vec{p}}{E}$$

The above given derivation of conservation laws can be directly extended to cover the case of a closed system which consists of free particles. In that case, one must introduce for each particle the action integral (5.25). In other words, one only must attach a particle index i to the variables and physical entities and sum (i=1,2,...,N). In this manner, invariance against translations analogously gives the conservation theorem of the total momentum,

$$\sum_i \vec{p}_i = \text{const} \tag{5.35}$$

and invariance against homogeneous Lorentz transformations gives the conservation of the angular momentum tensor ,

$$\sum_i \left(x_i^\mu p_i^\nu - x_i^\nu p_i^\mu \right) = \text{const} \tag{5.36}$$

While the spatial components of this tensor give the total angular momentum tensor, the mixed components yield the relativistic definition of the system's center-of-mass coordinates. Since the total energy ΣE_i is conserved, the relation

$$\sum_i \left(t \vec{p}_i - \frac{E_i}{c^2} \vec{x}_i \right) = \text{const}$$

and division by ΣE_i leads to

$$\frac{\Sigma E_i \vec{x}_i}{\Sigma E_i} = \frac{c^2 \Sigma \vec{p}_i}{\Sigma E_i} t + \text{const}$$

From this equation one can see that the "center of mass of the energy",

$$\vec{X} = \frac{\Sigma E_i \vec{x}_i}{\Sigma E_i} \tag{5.37}$$

moves uniformly along a straight line with the velocity

$$\vec{V} = \frac{c^2 \Sigma \vec{p}_i}{\Sigma E_i}$$

Accordingly, conservation of the angular momentum tensor $L^{\mu\nu}$ summarizes the conservation of three-dimensional angular momentum and the center-of-mass theorem.

If the particles have small velocities ($v_i \ll c$) then, with $E \approx m_i c^2$, from (5.37) the Newtonian expression of the center of mass,

$$\vec{X}_{NR} = \frac{\Sigma m_i \vec{x}_i}{\Sigma m_i}$$

will follow. It must be noted that the components of the vector (5.37) are not the spatial components of a four-vector. Thus, the position of the center of mass in space depends on the choice of the inertial system. The condition $\Sigma \vec{p}_i = 0$ distinguishes the center-of-mass system already used in Section 5.4, because in this inertial system the center of mass of the particles is at rest, $\vec{V} = 0$.

When deriving equations (5.35)-(5.37) it was assumed that the particles are free. If there are interactions between them, then one must take into consideration the momentum and energy contributions of the interaction transmitting fields, both when discussing the energy-momentum conservation and also when defining the relativistic center of mass. The above given sums over the energies and momenta of the particles must then be amended with the appropriate integrals over the energy and momentum densities of the fields. The energy-momentum tensor of the electromagnetic field will be discussed in the next Chapter.

To conclude our detailed derivation of the conservation laws from the invariance of the action integral (or equivalently, of the Lagrangian) against symmetry transformations, we illuminate this

simple connection by the following consideration. Because of the homogeneity and isotropy of space and time, the Lagrangian of a closed
system will not change under an inhomogeneous Lorentz transformation;
i.e., the parameters ε^μ and $\varepsilon^{\mu\nu}$ of these transformations are cycli-
cal coordinates. Therefore, the corresponding generalized momenta

$$\frac{\delta S}{\delta \varepsilon^\mu} = - p_\mu \, , \quad \frac{\delta S}{\delta \varepsilon^{\mu\nu}} = \frac{1}{2} \left(x_\mu p_\nu - x_\nu p_\mu \right) \, ,$$

(which can be obtained from the variation of the action integral,
cf. p.138) are constant in time. These conserved quantities are precisely the momentum vector and the angular momentum tensor. Thus,
one can use the connection between symmetry transformations and conserved quantities for the definition of the corresponding physical
observables.

Chapter 6

ELECTRODYNAMICS: AN EXAMPLE OF A RELATIVISTIC FIELD THEORY

Interactions between particles are transmitted by fields which transfer energy and momentum and thus have an independent physical reality (see Section 5.3). In particular, this holds for the electromagnetic field, which describes interactions between electrically charged particles. Since electrodynamics obeys Einstein's principle of relativity, the basic equations of this theory should be formulated in covariant form, i.e. as relations between tensorial quantities in Minkowski space. However, classical electrodynamics (unlike Newtonian mechanics) does not require any alteration: it already is a relativistic theory. We convince ourselves of this fact in the introductory Section that follows, where we shall write out the theory in covariant form.

6.1 Wave Equations for the Electromagnetic Potentials

The relativistic invariance of electrodynamics is directly manifest from the equations that govern electromagnetic potentials:

$$\vec{\nabla}^2 \vec{A} - \frac{1}{c^2} \frac{\partial^2 \vec{A}}{\partial t^2} = - \frac{4\pi}{c} \vec{j}$$

$$\vec{\nabla}^2 \varphi - \frac{1}{c^2} \frac{\partial^2 \varphi}{\partial t^2} = - 4\pi \varrho \tag{6.1}$$

Here $\rho(x)$ and $\vec{j}(x)$ denote the densities of charge and current which obey the equation of continuity

$$\frac{\partial \varrho}{\partial t} + \vec{\nabla} \cdot \vec{j} = 0 \tag{6.2}$$

To start with, we recall the connection of these equations with the Maxwell equations.

Expressing the field strengths $\vec{E}(x)$ and $\vec{B}(x)$ by the potentials,

$$\vec{E} = -\vec{\nabla}\varphi - \frac{1}{c}\frac{\partial \vec{A}}{\partial t}$$

$$\vec{B} = \vec{\nabla} \times \vec{A} \tag{6.3}$$

the homogeneous Maxwell equations[1] which do not contain terms for charge and current,

$$\vec{\nabla} \times \vec{E} + \frac{1}{c}\frac{\partial B}{\partial t} = 0 \quad , \quad \vec{\nabla} \cdot \vec{B} = 0 \tag{6.4a}$$

are identically satisfied. The behaviour of φ and $\vec{A}$ on sources is determined by the inhomogeneous Maxwell equations

$$\vec{\nabla} \times \vec{B} - \frac{1}{c}\frac{\partial \vec{E}}{\partial t} = \frac{4\pi}{\cdot c}\vec{j} \quad , \quad \vec{\nabla} \cdot \vec{E} = 4\pi\varrho \tag{6.4b}$$

which can be written in terms of the potentials in the following manner:

$$\vec{\nabla}^2 \vec{A} - \frac{1}{c^2}\frac{\partial^2 \vec{A}}{\partial t^2} = -\frac{4\pi}{c}\vec{j} + \vec{\nabla}\left(\vec{\nabla}\cdot\vec{A} + \frac{1}{c}\frac{\partial \varphi}{\partial t}\right)$$

$$\vec{\nabla}^2 \varphi + \frac{1}{c}\frac{\partial}{\partial t}(\vec{\nabla}\cdot\vec{A}) = -4\pi\varrho$$

---

[1] We use Gaussian cgs system units. For comparison with the international systems of units (IS) the reader is referred to the relevant textbooks, for example, J.D. Jackson, Classical Electrodynamics, 2nd ed. (J. Wiley, New York, 1975).

Note that the vector potential $\vec{A}$ is fixed only up to an arbitrary scalar function $\lambda(x)$, because making the substitution

$$\vec{A} \rightarrow \vec{A}' = \vec{A} + \vec{\nabla}\lambda.(x)$$

the value of $\vec{B} = \vec{\nabla} \times \vec{A}$ does not change. So as to keep the electric field likewise unchanged, the scalar potential φ must be also transformed in an appropriate manner. i.e.

$$\varphi \rightarrow \varphi' = \varphi - \frac{1}{c}\frac{\partial\lambda}{\partial t}$$

This remaining freedom in the choice of the potentials can be used to decouple the above equations. Thus, one chooses the potentials so that the relation

$$\vec{\nabla}\cdot\vec{A} + \frac{1}{c}\frac{\partial\varphi}{\partial t} = 0 \tag{6.5}$$

is satisfied. With this "Lorentz condition" the equations for the potentials assume the form as given in (6.1); these are equivalent to the Maxwell equations.

The relativistic invariance of electrodynamics can now be ascertained by observing that equations (6.1) may be written in a covariant form, without any alterations. In order to see this, let us first consider the current density. According to experience, the charge of a particle is independent of the reference frame, that is, it is an invariant quantity. However, the charge density transforms as the temporal component of a four-vector. One convinces oneself of this by forming, with the use of the invariant quantity $\rho\, d^3x$, the four-vector

$$\varrho\, d^3x\, dx^\mu = \varrho\, d^3x\, dt\, \frac{dx^\mu}{dt}$$

Since the functional determinant of the Lorentz transformation (rotation) is 1, the four-dimensional volume element d^4x (and therefore also $d^3x\,dt = d^4x/c$) behaves as a scalar. Hence the coefficient of $d^3x\,dt$ in the above equation must be a vector,

$$j^\mu = \varrho \, \frac{dx^\mu}{dt} \tag{6.6}$$

This is the four-vector of current density and it has the components

$$j^\mu = (c\varrho, \vec{j}), \quad \vec{j} = \varrho \vec{v} \tag{6.7}$$

where $\vec{j}$ stands for the usual three-dimensional current density. Thus, on the right-hand side of equations (6.1) we have the components of the four-vector $-4\pi j^\mu/c$. Since the differential operator

$$\Box = \frac{1}{c^2} \frac{\partial^2}{\partial t^2} - \vec{\nabla}^2$$

is an invariant quantity (see p.89), the potentials φ and $\vec{A}$ must form a four-vector,

$$A^\mu = (\varphi, \vec{A}) \tag{6.8}$$

Using all this, we have the covariant form of the equations which determine the four-potential of electromagnetic field caused by the moving charges:

$$\Box A^\mu = \frac{4\pi}{c} j^\mu \tag{6.9}$$

Using the four-vector corresponding to the derivative (see p.89) we can now rewrite the Lorentz condition and the equation of continuity in a similarly covariant form. We have

$$\frac{\partial A^\mu}{\partial x^\mu} = 0 \tag{6.10}$$

and

$$\frac{\partial j^\mu}{\partial x^\mu} = 0 \tag{6.11}$$

respectively. In summary, the laws of electrodynamics obey Einstein's principle of relativity.

6.2 Lagrangian for a Charge in an External Field

Suppose a particle with charge e moves in a given external eletromagnetic field, that is, we neglect the reaction of the charge onto the field. The action integral which pertains to this situation is composed of the already known action for a free particle (equ. (5.25)), and an additional term which describes the interaction of the particle with the field. This term must contain both quantities that refer to the particle as well quantities that characterize the field. Moreover, the Lagrangian has the dimension of energy. Thus, the term which is added to the free Lagrangian represents the interaction energy of the charge e in the external field. For a point charge in the scalar potential field $\varphi(x)$ this is exactly the potential energy $e\varphi$. With the charge density $\rho(x)$ one obtains the corresponding energy density $\rho\varphi$. Now, the interaction must be invariant against Lorentz transformations. Starting with $\rho\varphi$, this can be easily achieved by completing ρ and φ according to (6.7) and (6.8) to the four-vectors j_μ and A^μ (four-potential), respectively, and by forming the invariant product $j_\mu A^\mu/c$. (Since $j_0 = c\rho$, the factor $1/c$ must be used.) Integrating over this expression of the interaction energy density we obtain the action integral for the charge and current distribution j_μ in the external field A^μ:

$$S = - mc \int ds - \frac{1}{c} \int j_\mu A^\mu d^3x \, dt \qquad (6.12)$$

Here we have a negative sign in front of the contribution from the interaction, because in the limiting case of small velocities one has L=T-U, i.e., one must subtract from the Lagrangian of the free particle the potential energy of the charge in the potential field φ.

For a point charge, the charge density occurring in the definition (6.6) of the current is proportional to the delta function, and spatial integration gives the interaction term

$$S_w = - \frac{e}{c} \int A_\mu \frac{dx^\mu}{dt} \, dt$$

Thus, the complete action integral can now be written as

$$S = - \int \left(m c \, ds + \frac{e}{c} A_\mu dx^\mu \right) , \tag{6.13}$$

or, using $\vec{v} = d\vec{x}/dt$,

$$S = \int \left(- m c^2 \sqrt{1 - \frac{v^2}{c^2}} + \frac{e}{c} \vec{A} \cdot \vec{v} - e \varphi \right) dt$$

The Lagrangian, written in terms of the ordinary velocity,

$$\tilde{L} = - m c^2 \sqrt{1 - \frac{v^2}{c^2}} + \frac{e}{c} \vec{A} \cdot \vec{v} - e \varphi$$

differs from the Lagrangian of a free particle by the additional term $e \vec{A} \cdot \vec{v}/c - e\varphi$ which describes the interaction between the point charge and the electromagnetic field. This is a generalized potential which depends not only on the position but also on the velocity of the particle.

The above Lagrangian yields the generalized momentum

$$\vec{P} = \frac{\partial \tilde{L}}{\partial \vec{v}} = \vec{p} + \frac{e}{c} \vec{A} ,$$

where $\vec{p} = m\vec{u}$. To obtain the Hamiltonian, one must express

$$H = \vec{v} \cdot \frac{\partial \tilde{L}}{\partial \vec{v}} - \tilde{L} = \frac{m c^2}{\sqrt{1 - \frac{v^2}{c^2}}} + e \varphi$$

in terms of the generalized momentum $\vec{P}$. If one isolates in these relations $\vec{p}$ and $E = mc^2/\sqrt{1-\beta^2}$, then, because of the connection (5.12) between E and $\vec{p}$ one obtains the Hamiltonian in the form

$$H = c \sqrt{\left(\vec{P} - \frac{e}{c} \vec{A} \right)^2 + m^2 c^2} + e \varphi$$

Thus, the transition from a free particle to the case of interaction in an external field is achieved by replacing the energy and momentum of the free particle by $H - e\varphi$ and $\vec{P} - e\vec{A}/c$, respectively.

6.3 The Electromagnetic Field Tensor and the Equations of Motion in an External Field

We now want to derive from the invariant action integral (6.13) the equations of motion of a point particle in an external field. Using the parameter λ as done in Section 5.5, the action integral can be written in this form:

$$S = -\int \left\{ mc \sqrt{g_{\mu\nu} \frac{dx^{\mu}}{d\lambda} \frac{dx^{\nu}}{d\lambda}} + \frac{e}{c} \frac{dx^{\mu}}{d\lambda} A_{\mu}(x) \right\} d\lambda$$

From this one obtains the Euler-Lagrange equations (5.28)

$$mc \frac{d}{d\lambda} \left[\frac{\dot{x}_{\mu}}{\sqrt{g_{\mu\nu} \dot{x}^{\mu} \dot{x}^{\nu}}} \right] + \frac{e}{c} \frac{dA_{\mu}(x)}{d\lambda} - \frac{e}{c} \frac{dx^{\nu}}{d\lambda} \frac{\partial A_{\nu}(x)}{\partial x^{\mu}} = 0$$

and setting $\lambda = \tau$ as well as using equation (5.27), we have

$$m \frac{du_{\mu}}{d\tau} + \frac{e}{c} \frac{dA_{\mu}}{d\tau} - \frac{e}{c} \frac{dx^{\nu}}{d\tau} \frac{\partial A_{\nu}}{\partial x^{\mu}} = 0$$

Since

$$\frac{dA_{\mu}}{d\tau} = \frac{\partial A_{\mu}}{\partial x^{\nu}} \frac{dx^{\nu}}{d\tau}$$

we get herefrom

$$m \frac{du_{\mu}}{d\tau} = \frac{e}{c} \left(\frac{\partial A_{\nu}}{\partial x^{\mu}} - \frac{\partial A_{\mu}}{\partial x^{\nu}} \right) u^{\nu}; \quad u^{\nu} = \frac{dx^{\nu}}{d\tau}$$

or

$$\frac{dp_{\mu}}{d\tau} = \frac{e}{c} F_{\mu\nu} u^{\nu} \tag{6.14}$$

This is the covariant equation of motion of a point charge in an external field. The antisymmetric tensor

$$F_{\mu\nu} = \frac{\partial A_{\nu}}{\partial x^{\mu}} - \frac{\partial A_{\mu}}{\partial x^{\nu}} = \partial_{\mu} A_{\nu} - \partial_{\nu} A_{\mu} \tag{6.15}$$

which occurs here is called the electromagnetic field-strength tensor (or field tensor). Because of its antisymmetry, it has only six non-vanishing independent components.

We can convince ourselves that the field tensor (6.15) comprises in a covariant way the three-dimensional field strength $\vec{E}$ and $\vec{B}$. Using $A_\mu = (\varphi, -\vec{A})$, $\vec{A} = (A_x, A_y, A_z)$ and noting that $\vec{\nabla} \times \vec{A} = \vec{B}$, the spatial components F_{mn} become

$$F_{12} = \frac{\partial A_2}{\partial x^1} - \frac{\partial A_1}{\partial x^2} = -\left(\frac{\partial A_y}{\partial x} - \frac{\partial A_x}{\partial y} \right) = -B_z$$

and similarly

$$F_{13} = B_y$$

$$F_{23} = -B_x$$

Analogously we get for the space-time components F_{on}, if we use the definition (6.3) of the electric field strength $\vec{E}$,

$$F_{01} = \frac{\partial A_1}{\partial x^0} - \frac{\partial A_0}{\partial x^1} = -\frac{\partial \varphi}{\partial x} - \frac{1}{c} \frac{\partial A_x}{\partial t} = E_x$$

and

$$F_{02} = E_y$$

$$F_{03} = E_z$$

Finally, taking into account the antisymmetry of the tensor, $F_{\mu\nu} = -F_{\nu\mu}$, our results can be summarized in the following form:

$$F_{\mu\nu} = \begin{pmatrix} 0 & E_x & E_y & E_z \\ -E_x & 0 & -B_z & B_y \\ -E_y & B_z & 0 & -B_x \\ -E_z & -B_y & B_x & 0 \end{pmatrix} , \quad F^{\mu\nu} = \begin{pmatrix} 0 & -E_x & -E_y & -E_z \\ E_x & 0 & -B_z & B_y \\ E_y & B_z & 0 & -B_x \\ E_z & -B_y & B_x & 0 \end{pmatrix} \quad (6.16)$$

Here μ and ν refer to the row and column indices.

Now we consider the spatial components of the equation of motion (6.14). This can be written in terms of the contravariant components of $F^{\mu\nu}$ in the form

$$\frac{d p^\mu}{d\tau} = \frac{e}{c} F^{\mu\nu} u_\nu \quad (6.17)$$

With the four-velocity $u_\nu = (\gamma c, -\gamma \vec{v})$ and equation (6.16) this can be brought to the following form:

$$\frac{dp^1}{dt} = e\, E_x + \frac{e}{c} \left[\vec{v} \times \vec{B} \right]_x \quad,$$

and similarly for the other components. In this manner one obtains the equation of motion of a charge in the electromagnetic field, expressed in terms of the relevant three-vectors:

$$\frac{d\vec{p}}{dt} = e\vec{E} + \frac{e}{c} \left[\vec{v} \times \vec{B} \right] \tag{6.18}$$

where on the right-hand side we have the well-known Lorentz force that depends on the particle's velocity. It should be observed that $\vec{p} = \gamma m \vec{v}$ stands for the spatial part of the four-momentum which goes over into the classical expression $m\vec{v}$ only if $v \ll c$.

The temporal component in equation (6.17) gives

$$\gamma \frac{dp^0}{dt} = \frac{e}{c}\, F^{0\nu} u_\nu = \frac{e}{c}\, \gamma\, \vec{E} \cdot \vec{v}$$

i.e., using $p^0 = E/c$,

$$\frac{dE_{kin}}{dt} = e\, \vec{E} \cdot \vec{v}$$

because for the time derivative the constant rest energy is irrelevant. This equation tells us that the temporal charge of the kinetic energy equals the work $e\vec{E} \cdot \vec{v}$ done on the particle per unit time (cf.(5.9), p.109). Clearly, this work is done by the electric part $e\vec{E}$ of the force (6.18) alone. The magnetic field $\vec{B}$ does not perform work. The reason for this is that the force $e\vec{v} \times \vec{B}/c$ is always perpendicular to the velocity $\vec{v}$, i.e., the well-known relation $\vec{v} \cdot \vec{v} \times \vec{B} = 0$ pervails.

6.4 Transformation of the Field Components and the Invariants of the Field

Under a Lorentz transformation the electromagnetic potentials behave as a vector field (cf. equ.(3.7)),

$$A'^{\mu}(x') = \Lambda^{\mu}{}_{\alpha} A^{\alpha}(x)$$

On the other hand, the electromagnetic field strengths $\vec{E}$ and $\vec{B}$ transform as the components of the antisymmetric field tensor $F^{\mu\nu}$,

$$F'^{\mu\nu}(x') = \Lambda^{\mu}{}_{\alpha} \Lambda^{\nu}{}_{\beta} F^{\alpha\beta}(x) \tag{6.19}$$

We now work out the transformation formulae of the field strengths for a special Lorentz transformation $L^{\mu}{}_{\alpha}$. The non-zero elements of the transformation matrix $L^{\mu}{}_{\alpha}$ are

$$L^{0}{}_{0} = L^{1}{}_{1} = \gamma \ , \quad L^{0}{}_{1} = L^{1}{}_{0} = -\frac{v}{c}\gamma \ , \quad L^{2}{}_{2} = L^{3}{}_{3} = 1$$

(see p.92). According to (6.19) we have, using these elements,

$$F'^{10} = L^{1}{}_{0} L^{0}{}_{1} F^{01} + L^{1}{}_{1} L^{0}{}_{0} F^{10}$$

$$= \gamma^{2} \left(1 - \frac{v^{2}}{c^{2}}\right) F^{10} = F^{10}$$

i.e.,

$$E'_{x} = E_{x}$$

Since a Lorentz transformation with velocity v in the direction of the x^{1} axis does not change the coordinates x^{2} and x^{3}, the tensor component F^{32} remains unchanged; further, the components F^{20}, F^{30} transform as x^{0}, and F^{21}, F^{13} as x^{1}. In consequence, identifying the components (recall $F^{\mu\nu} = -F^{\nu\mu}$) one finally has

$$E'_{x} = E_{x} \ , \quad E'_{y} = \gamma\left(E_{y} - \frac{v}{c} B_{z}\right) \ , \quad E'_{z} = \gamma\left(E_{z} + \frac{v}{c} B_{y}\right)$$

$$B'_{x} = B_{x} \ , \quad B'_{y} = \gamma\left(B_{y} + \frac{v}{c} E_{z}\right) \ , \quad B'_{z} = \gamma\left(B_{z} - \frac{v}{c} E_{y}\right) \tag{6.20}$$

The inverse formulae are obtained as earlier (see p.24) by means of a relativistic interchange. These relations connect the numerical values of the field strengths at the world point x^{μ} referred to the system K with their values at x'^{μ} referred to K', where $x'^{\mu} = L^{\mu}{}_{\nu} x^{\nu}$. The arguments x' and x in the above transformation formulae still need amending. In this manner, in (6.20) the field strengths in the new coordinate sytem K' are expressed as functions of the old coordinates x. Substituting $x=L^{-1}x'$ on the right-hand sides of (6.20), one finally obtains the field strengths in K' as functions of the new coordinates x'.

Following the considerations on p.33, the above result can be generalized easily for the case when the new system K' moves relative to K with a velocity $\vec{v}$ pointing into an arbitrary direction. In this case, using the decomposition $\vec{E}_{\parallel} \parallel \vec{v}$, $\vec{E}_{\perp} \perp \vec{v}$ and similarly for $\vec{B}$, one obtains

$$\vec{E}'_{\parallel} = \vec{E}_{\parallel} \quad , \quad \vec{B}'_{\parallel} = \vec{B}_{\parallel}$$

$$\vec{E}'_{\perp} = \gamma \left(\vec{E}_{\perp} + \frac{1}{c} \vec{v} \times \vec{B} \right) \quad , \quad \vec{B}'_{\perp} = \gamma \left(\vec{B}_{\perp} - \frac{1}{c} \vec{v} \times \vec{E} \right)$$

Using the projection $\vec{E}_{\parallel} = \vec{v}(\vec{v} \cdot \vec{E})/v^2$ and $\vec{E} = \vec{E}_{\parallel} + \vec{E}_{\perp}$ as well as the corresponding relations for $\vec{B}$, one can rewrite this in the compact form

$$\vec{E}' = \gamma \left(\vec{E} + \frac{1}{c} \vec{v} \times \vec{B} \right) - (\gamma - 1) \frac{\vec{v} \cdot \vec{E}}{v^2} \vec{v}$$

$$\vec{B}' = \gamma \left(\vec{B} - \frac{\vec{v}}{c} \times \vec{E} \right) - (\gamma - 1) \frac{\vec{v} \cdot \vec{B}}{v^2} \vec{v}$$

(6.21)

For $v \ll c$ there follow herefrom the simpler relations

$$\vec{E}' = \vec{E} + \frac{1}{c} [\vec{v} \times \vec{B}] \quad , \quad \vec{B}' = \vec{B} - \frac{1}{c} [\vec{v} \times \vec{B}]$$

where contributions of the order v^2/c^2 are neglected.

From the transformation formulae of the field strengths we realize that statements such as "the field is purely electric (or magnetic)" have only a relative meaning. Clearly, an electric or a magnetic field may be zero in one reference system but non-zero in another one. In the general case of moving reference systems one has both an electric and a magnetic field. Only the two together form a quantity, viz. the field tensor $F^{\mu\nu}$, which characterizes the electromagnetic field in all inertial systems. The components of this tensor become mixed together according to the transformation law (6.19) when one passes to another reference system.

We illuminate these general remarks on the following simple example. Let us take an electrically charged particle which is at rest in system K. The corresponding electric field is the Coulomb field $\vec{E} = e\vec{x}/|x|^3$ which possesses spherical symmetry. If the particle has no magnetic moment, we also have $\vec{B} = 0$. However, in another reference system K' which is in relative motion to K, one finds a different electromagnetic field. As it can be seen from the transformation formulae (6.20), there is in K' also a magnetic field present. With the Maxwell equations, this circumstance is usually interpreted as follows: The charge which appears to be in motion in system K' represents a current which produces the observed magnetic field. This statement describes the relative character of the field $\vec{B}$ in another language.

From the formulae (6.20) we get for the electric field

$$\vec{E}\,(\vec{x}') = \left(1 - \frac{v^2}{c^2}\right) \frac{e\,\vec{x}'}{|\vec{x}'|^3 \left(1 - \frac{v^2}{c^2}\sin^2\theta'\right)^{3/2}}$$

Here we already replaced the coordinates x by $L^{-1}x'$ and, in addition, we took in K' the time fixed by t'= 0. The angle θ' measures the inclination of the vector $\vec{x}$ (which points from 0 to the point of observation) toward the direction of motion $(\vec{v})$. We see that $\vec{E}'$ is still a radial field but it is no longer spherically symmetric. For $\theta'= \pi/2$, $|\vec{E}'|$ is greater by a factor γ than the original magnitude $|\vec{E}|$, for $\theta'= 0$ it is smaller by a factor $1/\gamma^2$. Expressed this in terms of field lines of force, we see that these lines corresponding to the field of the moving electron are squashed toward the meridian plane $\theta'= \pi/2$. The Coulomb field of the electron is pressed flat. This effect increases with the electron's velocity.

Having discussed the transformation properties of the field $F_{\mu\nu}$, it will be interesting to find its invariants. These invariants are obtained from the characteristic equation of $F_{\mu\nu}$:

$$|F_{\mu\nu} - \lambda g_{\mu\nu}| = \lambda^4 + \lambda^2\,(\vec{B}^2 - \vec{E}^2) - (\vec{E}\cdot\vec{B})^2 = 0$$

The coefficients $(\vec{E}^2-\vec{B}^2)$ and $(\vec{E}\cdot\vec{B})$ are invariants and can be expressed also in terms of the field tensor:

$$I_1 = \vec{B}^2 - \vec{E}^2 = \frac{1}{2} F_{\mu\nu}F^{\mu\nu}$$

$$I_2 = -(\vec{E}\cdot\vec{B}) = \frac{1}{8} \epsilon^{\mu\nu\varrho\sigma} F_{\mu\nu} F_{\varrho\sigma}$$

(6.22)

In this form their invariance is directly transparent. Here $\epsilon^{\mu\nu\rho\sigma}$ stands for the completely antisymmetric Levi–Civita symbol with four indices which we defined on p.70. Introducing the field tensor $\tilde{F}^{\mu\nu}$ which is the dual of $F^{\mu\nu}$ by writing

$$\tilde{F}^{\mu\nu} = \frac{1}{2} \epsilon^{\mu\nu\varrho\sigma} F_{\varrho\sigma} = \begin{pmatrix} 0 & -B_x & -B_y & -B_z \\ B_x & 0 & E_z & -E_y \\ B_y & -E_z & 0 & E_x \\ B_z & E_y & -E_x & 0 \end{pmatrix}$$

(6.23)

then the invariant I_2 can be written as

$$I_2 = \frac{1}{4} F_{\mu\nu} \tilde{F}^{\mu\nu}$$

As seen from the definition (6.23), $\tilde{F}^{\mu\nu}$ is obtained from the field tensor $F^{\mu\nu}$ by the substitutions $\vec{E} \to \vec{B}$ and $\vec{B} \to -\vec{E}$. The quantities I_1 and I_2 are called fundamental invariants because all other that do not vanish can be expressed in terms of I_1 and I_2.

The invariance of I_1 and I_2 leads to interesting statements. Some of these we mention at this point. For example, if in a certain inertial system the magnitudes of the field strengths are equal ($|\vec{E}|=|\vec{B}|$), so that $I_1= 0$, then this holds for all inertial systems. Similarly, the statement that $\vec{E}$ and $\vec{B}$ are orthogonal, i.e. that $I_2= 0$, is valid in all inertial systems. More generally, statements about the ratio of the field strengths, $|\vec{E}|>|\vec{B}|$ (or $|\vec{E}|<|\vec{B}|$) have general validity. The sign of I_2 determines whether the angle between $\vec{E}$ and $\vec{B}$ is acute or obtuse. Supposing that $I_2= 0$, so that $\vec{E} \perp \vec{B}$, but $I_1> 0$, then in all inertial systems one has $|\vec{B}|>|\vec{E}|$ and one can find a system in which the electric field $\vec{E}$ (but not $\vec{B}$) vanishes. Similarly, one can find an inertial system in which there is no magnetic field $\vec{B}$, provided $I_1 < 0$ and $I_2 = 0$ hold true. A special case arises when both invariants vanish. The $\vec{E}$ and $\vec{B}$ will be equal in magnitude and orthogonal to each other in all inertial systems, as is the case, for example, with electromagnetic waves.

6.5   Gauge Invariance

The equation of motion (6.17) of a charge contains the field strengths $F^{\mu\nu}$ rather than the potentials. On the other hand, if the potentials A_μ are known then definition (6.15) leads to $F_{\mu\nu}$. However, to one and the same field $F^{\mu\nu}$ there belong different potentials. This is so because choosing new potentials by the substitution

$$A_\mu(x) \rightarrow A'_\mu(x) = A_\mu(x) + \frac{\partial}{\partial x^\mu} \lambda(x) \tag{6.24}$$

where $\lambda(x)$ is an arbitrary function, the $F^{\mu\nu}$ (and hence the equation of motion (6.17)) will not change, since the mixed derivatives of $\lambda(x)$ occur with opposite signs. This is called gauge invariance. Under the gauge transformation (6.24) the field strengths remain unchanged. The implication for the potentials is that specifying the field strengths, the potentials are not uniquely determined. Because of this ambiguity it is possible to choose the potentials in such a way that they satisfy an arbitrary, suitably determined subsidiary condition.

This freedom will now be utilized to simplify the equations which govern the potentials. We recall (see Section 6.1) that the covariant form of these equations is

$$\Box A^\mu - \partial^\mu \partial_\nu A^\nu = \frac{4\pi}{c} j^\mu$$

After "re-gauging"

$$A'^\mu = A^\mu + \partial^\mu \lambda$$

and demanding the condition

$$\partial_\mu \partial^\mu \lambda = \Box \lambda = -\partial_\nu A^\nu$$

one obtains for the four-divergence of the new potentials

$$\partial_\mu A'^\mu = \partial_\mu A^\mu + \Box \lambda = 0$$

Thus, the gauge was so chosen that the new potentials A'^μ obey the the so-called Lorentz condition

$$\partial_\mu A'^\mu = 0$$

and satisfy the simpler wave equations

$$\Box A'^\mu = \frac{4\pi}{c} j^\mu$$

This covariant gauge has the advantage that it is valid in all inertial systems.

Even the Lorentz condition does not uniquely fix A'^μ. One can re-gauge without violating the Lorentz condition, i.e. make the substi-tution

$$A'^\mu \rightarrow A'^\mu + \partial^\mu \eta \quad ,$$

provided the function η satisfies the homogeneous wave equation, i.e.

$$\Box \eta = 0$$

Gauge invariance has a strong connection with charge conserva-tion. For if one couples in a gauge invariant way the electromagnetic field to a charged matter field $\psi(x)$, then it follows from gauge invariance (which now implies a re-gauging of the potentials as well as a corresponding phase change in the complex matter field function) that the current density formed from the matter field satisfies the the equation of continuity so that charge is conserved.

In this connection we make a remark about the Lagrangian for a charge in an external field (see p.148). Even though the equation of motion (6.14) that follows from it is gauge invariant, this does not hold for the Lagrangian which depends explicitly on the potentials A_μ . However, the additional term in the action integral (6.13) which would arise from a gauge transformation can be written as a total differential:

$$\frac{e}{c} \frac{\partial \lambda}{\partial x^\mu} dx^\mu = \frac{e}{c} d\lambda$$

Such a term does not contribute to the variation of the action integral and hence it does not change the equation of motion.

6.6   Covariant Form of the Maxwell Equations

The Maxwell equations (6.4) can be given a covariant form if if one uses the electromagnetic field-strength tensor $F_{\mu\nu}$. First, from the definiton

$$F_{\mu\nu} = \partial_\mu A_\nu - \partial_\nu A_\mu$$

it directly follows that

$$\frac{\partial F_{23}}{\partial x^1} + \frac{\partial F_{31}}{\partial x^2} + \frac{\partial F_{12}}{\partial x^3} = 0 \tag{6.25}$$

On account of the antisymmetry of $F_{\mu\nu}$ the left-hand side of this equation is zero whenever any two indices (or all three indices) are equal. The relation (6.25) is then identically satisfied (0=0). The equations which for $\lambda \neq \mu \neq \nu$ are not identically satisfied, yield precisely the homogeneous Maxwell equations (6.4a). So, for example taking $\lambda=1$, $\mu=2$, $\nu=3$ we get

$$\frac{\partial F_{\mu\nu}}{\partial x^\lambda} + \frac{\partial F_{\nu\lambda}}{\partial x^\mu} + \frac{\partial F_{\lambda\mu}}{\partial x^\nu} = 0$$

that is, with (6.16)

$$\vec{\nabla} \cdot \vec{B} = 0$$

Writing down the other non-zero elements, one arrives at the remaining Maxwell equations in (6.4a).

Using the dual field tensor $\tilde{F}^{\mu\nu}$, the homogeneous Maxwell equations can be written in a particularly pregnant form:

$$\frac{\partial}{\partial x^{\nu}}\, \tilde{F}^{\mu\nu} = 0 \tag{6.26}$$

This equation is equivalent to (6.25). The four not identically satisfied relations which arise from (6.26) are precisely the Maxwell equations (6.4a). Equation (6.26) follows directly from the definition of $\tilde{F}^{\mu\nu}$ (and thus from $F^{\mu\nu}$), because, due to the interchangeability of the derivatives $\partial_{\nu}\partial_{\rho}$ and the antisymmetry of $\varepsilon^{\mu\nu\rho\sigma}$ we have

$$\partial_{\nu}\tilde{F}^{\mu\nu} = \frac{1}{2}\, \varepsilon^{\mu\nu\varrho\sigma}\, \partial_{\nu}\,(\partial_{\varrho}A_{\sigma} - \partial_{\sigma}A_{\varrho}) = \varepsilon^{\mu\nu\varrho\sigma}\,\partial_{\nu}\partial_{\varrho}A_{\sigma} = 0$$

Thus, the homogeneous Maxwell equations follow from the definition of the field tensor $F_{\mu\nu}$.

Before discussing in Section 6.8 how the inhomogeneous Maxwell equations, i.e. the field equations can be derived from the relativistic invariant action integral, we write them down in covariant form already at this point. Since in the Maxwell equations (6.4b) spatial and temporal derivatives of the fields $\vec{E}$ and $\vec{B}$ occur, and also because of the inhomogeneity connected with the charge and current density, it is plausible to make the following Ansatz:

$$\frac{\partial}{\partial x^{\nu}}\, F^{\mu\nu} = -\frac{4\pi}{c}\, j^{\mu} \tag{6.27}$$

With the definitions of $F^{\mu\nu}$ and j^{μ} one convinces oneself that (6.27) summarizes the equations (6.4b). In particular, for $\mu = 0$ one finds

$$\vec{\nabla} \cdot \vec{E} = 4\pi\varrho \quad,$$

and for $\mu = 1,2,3$ the other (vectoral) equation.

From the field equation (6.27) there follows directly the continuity equation (6.11) for the current density,

$$-\frac{4\pi}{c}\, \partial_{\mu}\, j^{\mu} = \partial_{\mu}\partial_{\nu}\, F^{\mu\nu} = 0 \quad,$$

because $F^{\mu\nu} = -F^{\nu\mu}$ is contracted with the symmetric tensor $\partial_\mu \partial_\nu$. If one introduces in (6.27) the potentials,

$$\partial^\mu \partial_\nu A^\nu - \Box A^\mu = -\frac{4\pi}{c} j^\mu ,$$

then, with the Lorentz condition $\partial_\nu A^\nu = 0$ one obtains the already known wave equation for the potentials,

$$\Box A^\mu = \frac{4\pi}{c} j^\mu .$$

In their covariant form, the laws of electrodynamics become particularly clear.

The Maxwell equations exhibit the following remarkable symmetry:

$$\partial_\nu \tilde{F}^{\mu\nu} = 0 \quad , \quad \partial_\nu F^{\mu\nu} = -\frac{4\pi}{c} j^\mu \qquad (6.28)$$

This symmetry would be really perfect if there existed a "magnetic current density" j_m^μ which would serve as the source of the dual field tensor $\tilde{F}^{\mu\nu}$. However, so far the experimental search for magnetic charges (monopoles) was unsuccessful. Nevertheless, this open question retains its interest, because in quantum theory the so far not understood quantization of the electric charge may be derived from the existence of magnetic monopoles.[1]

6.7   The Doppler Effect

Among the solutions of the Maxwell equations we mention here only the monochromatic plane waves in charge-free space. A discussion

----

[1] This idea was brought up by Dirac: P.A.M. Dirac, Proc.Roy.Soc. A133, 60(1931); Phys. Rev. 74, 817(1948). From the by now very extensive literature on this fascinating topic we refer to E. Amaldi and N. Cabibbo, in Aspects of Quantum Theory, A. Salam and E.P. Wigner, Eds. (Cambridge University Press, Cambridge, 1972), p. 183. There one can find further bibliographical notes.

of these special solutions leads immediately to the relativistic Doppler effect.

Let us consider a monochromatic plane wave with wavelength λ,

$$A^\mu (x) = A^\mu_{(0)} e^{i(\omega t - \vec{k} \cdot \vec{x})}$$

where $A^\mu_{(0)}$ is a constant vector, while $\omega = 2\pi\nu$ denotes the angular frequency and $\vec{k}$ the wave vector. The latter points into the direction of wave propagation and has the magnitude $2\pi/\lambda$. Obviously, the phase ϕ of the wave that occurs in the exponent of the e-function must be invariant under Lorentz transformations leading from K to K':

$$\phi = \omega t - \vec{k} \cdot \vec{x} = \omega' t' - \vec{k}' \cdot \vec{x}' \qquad (6.29)$$

The statement that in a given world point P the phase has a certain value (such as that the wave has a maximum amplitude or goes through a zero) is independent of the reference frame. Thus we recognize that in equation (6.29) we have the inner product of $x^\mu = (ct, \vec{x})$ with the four-vector

$$k^\mu = \left(\frac{\omega}{c}, \vec{k} \right) \quad .$$

This follows also from the fact that the plane wave, being a solution of the wave equation, leads to the condition

$$\left(\frac{\omega}{c} \right)^2 - \vec{k}^2 = 0$$

Because of the invariance of the d'Alambertian operator $\square$, this must hold in all reference frames. We see that the four-vector $k^\mu = (\omega/c, \vec{k})$ is lightlike. This corresponds to the fact that electromagnetic waves propagate with the speed of light.

Now, the formula for the relativistic Doppler effect simply follows from the Lorentz transformation of the wave four-vector. Let K' be a reference system which moves relative to K in the positive x^1 direction with the velocity v. Let the wave four-vector have the components $k^\mu = (1, \cos\Theta, \sin\Theta, 0) \omega/c$ in K, and the components $k'^\mu = (1, \cos\Theta', \sin\Theta', 0) \omega'/c$ in K'. Using the transformation law

$$k'^{\mu} = L^{\mu}{}_{\nu} k^{\nu}$$

(where the matrix $L^{\mu}{}_{\nu}$ is given on p.92) and concentrating on the temporal components $k'^{0} = \omega'/c$, one finds the formula for the relativistic Doppler effect:

$$\omega' = \omega \, \frac{1 - \beta \cos \theta}{\sqrt{1 - \beta^2}} \tag{6.30}$$

The other two components, together with (6.30), yield the following relations between the angles θ and θ':

$$\cos \theta' = \frac{\cos \theta - \beta}{1 - \beta \cos \theta} \quad , \quad \sin \theta' = \frac{\sqrt{1 - \beta^2} \, \sin \theta}{1 - \beta \cos \theta} \tag{6.31}$$

This is the formula for the aberration (see also p.45).

The longitudinal Doppler effect (for instance with $\vec{k}$ in the direction of $\vec{v}$, i.e. with $\theta = \theta' = 0$) gives specifically

$$\omega' = \omega \sqrt{\frac{1 - \beta}{1 + \beta}} \approx \omega \, (1 - \beta) \tag{6.32}$$

which shows that this effect is of first order in β in the lowest approximation. In contrast, the transversal effect $(\theta = \pi/2)$ gives

$$\omega' = \frac{\omega}{\sqrt{1 - \beta^2}} \approx \omega \left(1 + \frac{1}{2} \beta^2 \right) \tag{6.33}$$

which reveals a quadratic dependence on β.

The transversal Doppler effect reproduces directly the time dilation of a moving clock. For, if one considers the emitter of the electromagnetic wave, which is at rest in K', as a signal source with a time interval $\Delta t_0 = \Delta t' = 2\pi/\omega'$, then this time interval is measured by the observer in K to have the value $\Delta t = 2\pi/\omega$, which is to say (see equ.(6.33))

$$\Delta t = \frac{\Delta t_0}{\sqrt{1 - \beta^2}} \quad , \quad \Delta t' = \Delta t_0$$

But this is precisely what time dilation means (see p.40, equ.(2.17)).

In this manner, the transversal Doppler effect is a typical relativistic effect. It does not occur in classical physics. Its direct verification[1] is a proof of time dilation and hence of special relativity theory.

We conclude this section with a brief remark on the duality of particles and waves which is a substantial ingredient of quantum theory. Following M. Planck, the quantum hypothesis for light (electromagnetic waves) is

$$E = \hbar \omega \quad , \quad \hbar = \frac{h}{2\pi} \tag{6.34}$$

Both the energy E of the light quantum as well as the angular frequency ω of the light wave are temporal components of four-vectors, viz. q^μ (momentum) and k^μ (wave number). Since the above relation must hold independently of the chosen inertial system, the Planck constant $\hbar$ is invariant against Lorentz transformations, as one would expect for a universal constant of nature. Furthermore, the completing spatial components $\vec{q}$ and $\hbar\vec{k}$ must coincide[2], so that the Planck condition for the energy results in the corresponding relation for the momentum of the light quantum, i.e.

$$\vec{q} = \hbar \vec{k} \quad , \quad |\vec{k}| = \frac{2\pi}{\lambda}$$

In covariant form the connection between the corpuscular entities $E, \vec{q}$ and the wave entities $\omega, \vec{k}$ read

$$q^\mu = \hbar k^\mu \tag{6.35}$$

The plausible extension of this relation between corpuscular and wave entities for the case of non-zero rest mass was eventually suggested by L. de Broglie. This idea which is consistent with special relativity

---

[1] Using the resonance absorption of γ-rays (Mössbauer effect), the transversal Doppler effect could be verified with an accuracy of a few percent: H. Hay, J. Schiffer, T. Cranshaw, and P. Egelstaff, Phys. Rev. Letters 4, 165 (1960).

[2] Here we used the following property of four-vectors: If two four-vectors have a particular one of their components equal in all inertial systems, then they are identical.

and in fact is suggested by it, i.e. the wavelike properties of material particles, contributed decisively to the development of quantum theory.

6.8 Action Integral for the Electromagnetic Field and the Field Equations

In Section 6.3 we were studying the equation of motion of a charge in a given external electromagnetic field A^μ. When deriving the equation of motion (6.14), only the coordinates of the charge were varied, the field itself stayed constant and hence did not experience a reaction. The action integral (6.13) consists of two parts, $S = S_M + S_W$, where S_M contains only the variables of the charged particle (of matter) and S_W describes the interaction of the particle with the external field. The action integral S_M taken by itself leads to the equation of motion of the free particle (see Section 5.5).

A contribution to the action integral which would contain only the electromagnetic field quantities, has not yet been considered. The corresponding action integral S_F would lead to the free equations of motion of the field, i.e. the free field equations, if one varied the field quantities which should be envisaged as the generalized coordinates. Thus, eventually, the action integral for the entire system consisting of field and charges (matter) is composed additively from the contributions S_F, S_M, and S_W,

$$S = S_F + S_M + S_W$$

and the particle and field quantities have to be varied independently. If the variation of S is done with respect to the particle coordinates, then again the equation of motion (6.14) is obtained. From the variation of S with respect to the field quantities there follow the equations of motion of the field in interaction with charges, i.e. the field equations for the case of a given interaction.

The form of S_F is determined to a large extent by the following conditions:

a) The action integral S_F must be invariant against Lorentz transformations.

b) S_F must be gauge invariant, i.e. it contains explicitly only the field strengths but not the potentials.[1]

c) The electromagnetic field obeys the superposition principle. This experiential property implies that the field equations are linear differential equations. The sum of two arbitrary solutions of the field equations is again a solution. Now, in order to get linear field equations when the variation of S_F is taken, the integrand of S_F must be quadratic in the field components.

These simple requirements suffice to find S_F. The first requirement can be satisfied by starting with the fundamental invariants I_1 and I_2. Since I_2 is a pseudo-scalar and the squares $(I_1)^2$ and $(I_2)^2$ would not lead to linear differential equations, one must use only I_1 itself. The Ansatz

$$S_F \sim \int F_{\mu\nu}\, F^{\mu\nu} d^3x\, dt$$

satisfies also the second requirement. One should note that an invariant term of the form $A_\mu A^\mu$ violates gauge invariance. In analogy to mechanics, one does not allow for higher derivatives of the variables, i.e. for derivatives of $F_{\mu\nu}$. The still missing numerical factor in front of the integral depends on the choice of the system of units, and it is fixed by the definition of the field strengths and the Maxwell equations. In the Gauss system used here, its value is $-1/16\pi$. The negative sign arises because, in analogy to mechanics, the "kinetic terms" $(\partial \vec{A}/\partial t)^2$ which, on account of $F_{\mu\nu}F^{\mu\nu}=2(\vec{B}^2-\vec{E}^2)$ occur in $\vec{E}^2$,

---

[1] This does not apply to the interaction term S_w. However, as we already saw (p.158), here a gauge transformation does not alter the equations of motion. - A better insight is obtained by the following consideration. Because of current conservation, the change in the integrand may be written as a divergence $\partial_\mu(j^\mu\lambda)$. According to the Gauss theorem, the four-dimensional volume integral of this term gives an integral over the boundary of the integration domain. But this does not contribute to the variation which is used to derive the equations of motion. This consideration illuminates the connection of gauge invariance with charge conservation.

are required to have a positive sign. Otherwise S_F had no minimum as required by the principle of least action.

Nevertheless, one should not have the impression that the integrand of S_F is now uniquely determined. Rather, it ought to be emphasized that this integrand is fixed only up to an additive divergence of a four-vector which may depend on the quantities to be varied. This is so because the surface integral that follows from such a term by the Gauss theorem does not contribute to the equations of motion since the variations vanish at the boundary of the volume. A relevant example of this can be found in the preceding footnote on p. 165.

In summary, we now have the action integral for a system of charges:

$$S = - \frac{1}{16\pi c} \int d^4x \; F_{\mu\nu} F^{\mu\nu} - \sum_i m_i c \int ds_i - \frac{1}{c^2} \int d^4x \; j^\mu A_\mu \qquad (6.36)$$

(Note that $d^4x = c\,dt\,dx^3$.) Here $j^\mu(x)$ is the total current density, and A_μ and $F_{\mu\nu}$ represent the total field which contains both the field produced by the charges and also a possible external field. The field produced by the charges depends on the positions and velocities of these charges.

If one wants to derive the field equations, one must take the motion of the charges as given and take the variation of only the potentials (field functions). Since S_M depends only on the particle variables, variation with respect to the potentials gives the field equations via

$$\delta S = \delta (S_F + S_W) = 0$$

Thus, one must construct from this the Euler-Lagrange equations where the A_μ figure as "generalized coordinates". This can be done for fields ("field mechanics") in analogy to point mechanics.

The following tabulation displays the most important corresponding quantities:[1]

	Particle	Field
State variables (generalized coordinates)	$q_i(t)$	$\varphi(\vec{x}, t)$
Independent variables	i, t	$\vec{x}, t$
Summation over	i	$\vec{x}$
Lagrangian	$L = \sum_i L_i(q_i, \dot{q}_i)$	$L = \int \mathcal{L}(\varphi, \partial_\mu \varphi) \, d^3x$
Action integral	$\int L \, dt$	$\int \mathcal{L} \, d^3x \, dt$
Lagrangian derivatives	$[L]_i := \dfrac{\partial L}{\partial q_i} - \dfrac{d}{dt} \dfrac{\partial L}{\partial \dot{q}_i}$	$[\mathcal{L}]_\varphi := \dfrac{\partial \mathcal{L}}{\partial \varphi} - \dfrac{\partial}{\partial x^\mu} \dfrac{\partial \mathcal{L}}{\partial \partial_\mu \varphi}$

To simplify writing, we suppressed component indices of the field $\varphi(x)$. One observes that, for example, the four-gradient $\partial_\mu \varphi$ corresponds to the particle velocities $\dot{q}_i$. The integrand $\mathcal{L}(\varphi, \partial_\mu \varphi)$ of the action integral transforms as a scalar density and is therefore called the Lagrangian density.

Now, in our case, the Lagrangian density corresponding to the action integral $S_F + S_W$ is

$$\mathcal{L} = -\frac{1}{16\pi} F_{\alpha\beta} F^{\alpha\beta} - \frac{1}{c} j^\mu A_\mu \tag{6.37}$$

From the variation principle it follows that the Lagrangian derivatives vanish:

$$[\mathcal{L}]^\mu := \frac{\partial \mathcal{L}}{\partial A_\mu} - \partial_\nu \frac{\partial \mathcal{L}}{\partial \partial_\nu A_\mu} = 0 \tag{6.38}$$

---

[1] A detailed discussion can be found in H. Goldstein, Classical Mechanics, 2nd ed. (Addison-Wesley, Reading, Mass., 1980).

These are nothing else but the Euler-Lagrange equations. One has immediately

$$\frac{\partial \mathcal{L}}{\partial A_\mu} = -\frac{1}{c} j^\mu$$

If $F_{\mu\nu}$ is expressed in terms of the potentials and one forms the derivatives while noting the antisymmetry $F_{\mu\nu} = -F_{\nu\mu}$, one finds that

$$\frac{\partial \mathcal{L}}{\partial \partial_\nu A_\mu} = \frac{1}{4\pi} F^{\mu\nu}$$

Now the field equations follow:

$$\frac{\partial}{\partial x^\nu} F^{\mu\nu} = -\frac{4\pi}{c} j^\mu \tag{6.39}$$

As we already saw in Section 6.6, these summarize the inhomogeneous Maxwell equations (6.4b) in covariant form.

6.9 Noether's Theorem

The conservation laws of energy, momentum, and angular momentum are also valid for closed systems in electrodynamics. (A system is closed if it is not acted upon by external forces.) These conservation laws follow from the invariance of the dynamical laws with respect to temporal and spatial translations and homogeneous Lorentz transformations. Mathematically, the relation between the symmetry transformations and the conservation laws is characterized by Noether's Theorem,[1] which also permits the explicit calculation of the conserved entities. We now give a brief derivation of this important theorem.[2] In this manner, the method described in Section 5.6 will

---

[1] E. Noether, Nachr. Akad. Wiss. Göttingen, Math.-Phys. Kl. 235(1918).

[2] A more detailed derivation and bibliographical references can be found in U.E. Schröder, Fortschr. d. Physik 16, 357(1968); reprinted in: Symmetry in Physics, Selected Reprints, J. Rosen, Ed. (American Assoc. of Physics Teachers, New York, 1982). This Volume also contains an extensive bibliography on the topic.

be extended to field theories.

Let us consider the action integral[1] of the Lagrange density $\mathcal{L}$:

$$S = \int d^4x \, \mathcal{L}(\psi, \partial_\mu \psi, x^\mu)$$

In general, the field functions ("fields") $\psi(x)$ may be complex, and they may have the transformation character of any tensor- or spinor field. However, in this discussion we suppress the relevant indices.

The action integral S must be unchanged under a symmetry transformation. Ortherwise, the ensuing equations of motion would not be covariant. Keeping this in mind, we now wish to determine the formal change δS of the action integral that results from infinitesimal coordinate transformation

$$x^\mu \longrightarrow x'^\mu = x^\mu + \delta x^\mu$$

Here, the δx^μ include the infinitesimal parameters of the transformation and they may also depend on the coordinates. (As an example, the reader should think of infinitesimal Lorentz transformations introduced earlier in Section 5.6.) The transformation of the field functions induced by the infinitesimal coordinate transformation is of the form

$$\psi \rightarrow \psi'(x') = \psi(x) + \delta \psi(x)$$

and similarly we have for the derivatives:

$$\frac{\partial \psi}{\partial x^\mu} \longrightarrow \frac{\partial \psi(x')}{\partial x'^\mu} = \frac{\partial \psi(x)}{\partial x^\mu} + \delta \frac{\partial \psi(x)}{\partial x^\mu}$$

In these expressions the $\delta \psi(x)$ (and the $\delta \partial_\mu \psi$) denote the infinitesimal changes of the fields, which were caused by the coordinate transformation. Despite the customary use of the symbol δ, these changes must

---

[1] We permit an explicit coordinate dependence, even though the Lagrange density function $\mathcal{L}_F$ of the electromagnetic field does not have this feature. Allowing for explicit coordinate dependence permits the discussion of systems that are subject to prescribed external (spatially and temporally varying) effects.

not be confused with the variations of the path that are used in the derivation of the equations of motion. The "total variations" we now consider do not commute with the operation of taking a partial derivative, i.e. $\delta(\partial_\mu\psi)\neq\partial_\mu(\delta\psi)$. On the other hand, it is advantageous to define a "local variation" $\overline{\delta}\psi$ by the equation

$$\psi'(x) = \psi(x) + \overline{\delta}\psi(x)$$

because $\overline{\delta}$ and ∂_μ do commute. (Note that $\overline{\delta}\psi$ is defined with keeping arguments unchanged.) The relation between the two infinitesimal variations is given by

$$\delta\psi = \psi'(x+\delta x) - \psi(x) = \psi'(x) - \psi(x) + \frac{\partial\psi}{\partial x}\,\delta x$$

$$\delta\psi = \overline{\delta}\psi(x) + \frac{\partial\psi}{\partial x}\delta x \tag{6.40}$$

where we suppressed Lorentz indices.

Using the shorthand $\partial_\mu\psi = \psi,_\mu$, the change of the action integral can be written as

$$\delta S = \int d^4x\left\{\frac{\partial\mathcal{L}}{\partial\psi}\,\delta\psi + \frac{\partial\mathcal{L}}{\partial\psi,_\mu}\,\delta(\psi,_\mu) + \left(\frac{\partial\mathcal{L}}{\partial x^\mu}\right)_x\delta x^\mu\right\} + \int\mathcal{L}\,\delta(d^4x)\ ,$$

Here the $(\partial\mathcal{L}/\partial x^\mu)_x$ indicate partials of $\mathcal{L}$ only with respect to the potentially present explicit x^μ coordinate dependence of $\mathcal{L}$. The transformation of the volume element d^4x can be calculated by using the Jacobian, i.e.,

$$d^4x' = \left|\frac{\partial(x^\nu + \delta x^\nu)}{\partial x^\mu}\right|d^4x\ .$$

Explicit evaluation yields

$$d^4x' = \left(1 + \partial_\nu\delta x^\nu\right)d^4x + 0\left((\delta x)^2\right)$$

so that

$$\delta(d^4x) = \partial_\nu\delta x^\nu d^4x$$

Here, as in the sequel, second- and higher-order terms in δx are neglected.

Next, using equation (6.40), we introduce the local variation $\overline{\delta}\psi$

into the expression for δS and get

$$\delta S = \int d^4x \left\{ \frac{\partial \mathscr{L}}{\partial \psi} \bar{\delta} \psi + \frac{\partial \mathscr{L}}{\partial \psi_{,\mu}} \bar{\delta}(\psi_{,\mu}) + \frac{\partial \mathscr{L}}{\partial x^\mu} \delta x^\mu \right\} + \int d^4x \, \mathscr{L} \partial_\nu \delta x^\nu \, ,$$

where

$$\frac{\partial \mathscr{L}}{\partial x^\mu} \equiv \left(\frac{\partial \mathscr{L}}{\partial x^\mu} \right)_x + \frac{\partial \mathscr{L}}{\partial \psi} \frac{\partial \psi}{\partial x^\mu} + \frac{\partial \mathscr{L}}{\partial \psi_{,\nu}} \frac{\partial \psi_{,\nu}}{\partial x^\mu}$$

For convenience, we use the shorthand $\partial_\mu \mathscr{L}$ to denote $\partial \mathscr{L}/\partial x^\mu$. If we then note the commutativity $\bar{\delta}(\partial_\mu \psi) = \partial_\mu(\bar{\delta}\psi)$, we find

$$\delta S = \int d^4x \left\{ \frac{\partial \mathscr{L}}{\partial \psi} \bar{\delta} \psi + \frac{\partial \mathscr{L}}{\partial \psi_{,\mu}} \partial_\mu(\bar{\delta}\psi) + \partial_\mu(\mathscr{L} \delta x^\mu) \right\}$$

Partial integration of the second term yields

$$\int d^4x \frac{\partial \mathscr{L}}{\partial \psi_{,\mu}} \partial_\mu(\bar{\delta}\psi) = \oint d\sigma^\mu \frac{\partial \mathscr{L}}{\partial \psi_{,\mu}} \bar{\delta}\psi - \int d^4x \, \partial_\mu \frac{\partial \mathscr{L}}{\partial \psi_{,\mu}} \bar{\delta}\psi \, .$$

The ensuing surface integral may be transformed by Gauss's theorem. Recalling that, for the assumed symmetry transformations, δS must be zero, we finally obtain

$$\delta S = \int d^4x \left\{ [\mathscr{L}]_\psi \bar{\delta}\psi + \partial_\mu \left(\frac{\partial \mathscr{L}}{\partial \psi_{,\mu}} \bar{\delta}\psi + \mathscr{L} g^{\mu\nu} \delta x_\nu \right) \right\} = 0$$

But, since the domain of integration is arbitrary, this implies

$$[\mathscr{L}]_\psi \bar{\delta}\psi + \partial_\mu \left\{ \frac{\partial \mathscr{L}}{\partial \psi_{,\mu}} \bar{\delta}\psi + \mathscr{L} g^{\mu\nu} \delta x_\nu \right\} = 0 \qquad (6.41)$$

This equation can be taken as the mathematical statement of Noether's theorem.

The relevant physical content of the theorem can be extracted from the above formula by reference to Hamilton's variation principle and its connection with the Lagrangian derivatives. Clearly, the physically realizable fields obey the field equations. Hence, for a closed system one must set $[\mathscr{L}]_\psi = 0$ so that

$$\partial_\mu \left\{ \frac{\partial \mathcal{L}}{\partial \psi_{,\mu}} \, \delta\psi - \frac{\partial \mathcal{L}}{\partial \psi_{,\mu}} \, \partial^\nu \psi \, \delta x_\nu + \mathcal{L} g^{\mu\nu} \delta x_\nu \right\} = 0 \qquad (6.42)$$

where, using (6.40), we once again re-introduced the total variation of the field functions $\delta\psi$.

Defining a tensor $\Theta^{\mu\nu}$ by setting

$$\Theta^{\mu\nu} := \frac{\partial \mathcal{L}}{\partial \psi_{,\mu}} \, \frac{\partial \psi}{\partial x_\nu} - \mathcal{L} g^{\mu\nu} \qquad (6.43)$$

equation (6.42) can be recast into the compact form

$$\partial_\mu g^\mu = 0 \; ; \quad g^\mu = \frac{\partial \mathcal{L}}{\partial \psi_{,\mu}} \, \delta\psi - \Theta^{\mu\nu} \delta x_\nu \qquad (6.44)$$

This is an equation of continuity for the four-vector g^μ (called a "current density"). Spatial integration and Gauss's theorem gives

$$\frac{\partial}{\partial x^0} \int d^3x \, g^0 = -\oint d\vec{f} \cdot \vec{g} \qquad (6.45)$$

The surface integral will vanish provided that, when integrating over the entire space, the field functions and their derivatives vanish fast enough so that the "current density" $\vec{g}$ falls off faster than the surface increases. In this case, we have the conservation law in an integral form:

$$\frac{\partial}{\partial x^0} \int d^3x \, g^0 = 0 \; . \qquad (6.46)$$

That is, the integral of the time-component of the four-vector g^μ is independent of time. Similarly, as we saw earlier in the case of electric charge conservation, this integral is a scalar quantity.

If we integrate over a finite portion of volume V and hence, have a non-vanishing surface integral contribution, then (6.45) must be interpreted as an equation expressing a balance. The decrease per unit time of the total quantity (a "charge") inside the finite volume V equals the flux that, per unit time interval, passes outward through the surface boundary of V.

In summary, we can formulate the essential physical content of Noether's theorem as follows:

Assume that the action integral S formed from the Lagrangian density $\mathcal{L}(\psi, \partial\psi, x)$ is invariant with respect to a continuous transformation group G_n with an analytic dependence of its elements on n (independent) parameters. Then, the field equations $[\mathcal{L}]_\psi = 0$ imply n conservation laws:

$$\partial_\mu g^\mu_a = 0 \,, \qquad a = 1, 2, \ldots, n \,;$$

$$g^\mu_a \equiv \frac{\partial\mathcal{L}}{\partial\psi_{,\mu}} \, \delta_a\psi - \theta^{\mu\nu} \, \delta_a x_\nu \,, \tag{6.47}$$

Here $\Theta^{\mu\nu}$ stands for the tensor defined by (6.43).

As already noted, the field equations remain unchanged if one modifies the Lagrangian density by the addition of a divergence, i.e. if

$$\mathcal{L} \rightarrow \mathcal{L} + \partial_\mu \Omega^\mu (x, \psi) \,.$$

Allowing for this freedom in the discussion of the consequence of a symmetry group, one finds that Noether's theorem still follows. However, in the definition of the conserved current thus obtained, an additional term, $\delta\Omega^\mu$, will appear. Now, by appropriately choosing such additional divergence terms it can be arranged that the resulting Lagrangian density is exactly invariant against the transformations under consideration. In the following we shall assume that this exact invariance holds.

We illuminate the use of Noether's theorem by two examples.

1. Translations

Let the action be invariant with respect to the infinitesimal translations

$$x^\mu \rightarrow x'^\mu = x^\mu + \delta\epsilon^\mu \,, \qquad \delta\epsilon^\mu = \text{const} \,.$$

We have

$$\psi(x) = \psi\big(x(x')\big) = \psi'(x')$$

and accordingly, after introducing the parameters of translation,

$$\psi'(x) = \psi(x - \delta\epsilon)$$

Therefore, the definition (6.40) tells us that $\delta\psi = 0$. Hence from (6.44),

$$\partial_\mu \theta^{\mu\nu} \delta\epsilon_\nu = 0$$

Since the four parameters of the translation $\delta\epsilon_\nu$ are independent, we obtain the four conservation laws

$$\partial_\mu \theta^{\mu\nu} = 0 \quad . \tag{6.48}$$

The integral quantity

$$P^\mu = \frac{1}{c} \int d^3x \; \theta^{0\mu}$$

is a constant in time. In particular, we recognize that the integrand of

$$P^0 = \frac{1}{c} \int d^3x \left\{ \frac{\partial \mathcal{L}}{\partial \dot{\psi}} \dot{\psi} - \mathcal{L} \right\} \quad ,$$

is the familiar Hamiltonian density,

$$\mathcal{H} = \theta^{00} \quad ,$$

Hence, we can identify P^μ with the conserved energy-momentum vector of the field. The quantity $\Theta^{\mu\nu}$ is called the canonical energy-momentum tensor of the system.

The conserved field angular momentum can be obtained in an analogous manner. For this calculation one must specify the behaviour of the field ψ under Lorentz transformations. The relevant change $\delta\psi$ is zero only for a scalar field.

2. Phase transformations

For complex field functions, a real action integral must be invariant under the phase transformation

$$\psi \rightarrow \psi'(x) = e^{i\alpha} \psi(x) \quad ,$$

where α is a constant (and real) parameter. Since the coordinates are not transformed, we now have $\delta x^\mu = 0$. For the infinitesimal variations of the field one obtains

$$\delta\psi = i\,\delta\alpha\,\psi \quad, \quad \delta\psi^* = -i\,\delta\alpha\,\psi^*$$

and consequently (6.44) leads to the equation of continuity:

$$\partial_\mu\, j^\mu = 0 \quad, \quad j^\mu = i\alpha\left(\frac{\partial\mathcal{L}}{\partial\psi_{,\mu}}\,\psi - \frac{\partial\mathcal{L}}{\partial\psi^*_{,\mu}}\,\psi^*\right)$$

If the matter field ψ is coupled to the electromagnetic field, then j^μ is the current density associated with the electric charge:

$$Q = \int d^3x\, j^0(x)$$

Noether's theorem guarantees its conservation.

The above results can be easily generalized for other internal symmetry groups such as SU(2), SU(3), etc. These groups play an important role in the theory of elementary particles.

6.10 Energy-Momentum Tensor of the Electromagnetic Field

We now proceed to apply Noether's theorem to the free electromagnetic field and to derive in this context the energy-momentum tensor. The action integral formed with the Lagrangian density

$$\mathcal{L}_F = -\frac{1}{16\pi}\, F_{\mu\nu}\, F^{\mu\nu}$$

is invariant against translations $\delta x^\mu = \delta\varepsilon^\mu$. As we already saw in the previous example, for translations $\delta A^\mu = 0$. Hence, Noether's theorem (6.47) gives

$$\partial_\mu\, \theta^{\mu\nu} = 0 \tag{6.49}$$

Here

$$\theta^{\mu\nu} = \frac{\partial\mathcal{L}_F}{\partial A_{\lambda,\mu}}\, \partial^\nu A_\lambda - g^{\mu\nu}\, \mathcal{L}_F$$

is the canonical tensor of the electromagnetic field. Using

$$\frac{\partial \mathcal{L}_F}{\partial A_{\lambda,\mu}} = -\frac{1}{4\pi} F^{\mu\lambda} \qquad (6.50)$$

one obtains

$$\theta^{\mu\nu} = -\frac{1}{4\pi} F^{\mu\lambda} \partial^\nu A_\lambda - g^{\mu\nu} \mathcal{L}_F \qquad (6.51)$$

In view of (6.49),

$$P^\nu = \frac{1}{c} \int d^3x \; \theta^{o\nu} \qquad (6.52)$$

gives a conserved four-vector which we identify with the momentum of the field. The factor $1/c$ has the effect that, as earlier,

$$P^\nu = \left(\frac{W}{c}, \vec{P} \right)$$

where, in accord with the example given in the preceding section, W denotes the total field energy.

The momentum P^ν may also be written as an integral of $\theta^{\mu\nu}$ over an arbitrary spacelike hyperplane σ.[1] To see this, let us integrate $\partial_\mu \theta^{\mu\nu}$ over an arbitrary four-dimensional domain Γ which is bounded by two spacelike hyperplanes σ_1 and σ_2 and a side surface F which lies in spatial infinity. Due to the Gauss theorem, the volume integral can be decomposed into contributions from the surfaces as follows:

$$\int_\Gamma d^4x \; \partial_\mu \theta^{\mu\nu} = \int_{\sigma_2} d\sigma_\mu \theta^{\mu\nu} - \int_{\sigma_1} d\sigma_\mu \theta^{\mu\nu} + \int_F d\sigma_\mu \theta^{\mu\nu}$$

(We inverted here the sign of the surface normal of σ_1.) Since there must not be a field at spatial infinity, the integral over F vanishes. Thus, if equation (6.49) holds, we have

$$\int_{\sigma_2} d\sigma_\mu \theta^{\mu\nu} = \int_{\sigma_1} d\sigma_\mu \theta^{\mu\nu}$$

---

[1] A hyperplane is called spacelike if the normal vector to the surface (and hence $d\sigma_\mu$) is everywhere timelike. In that case the separation of two world points on the surface is spacelike.

We see that the integral of the divergenceless tensor $\Theta^{\mu\nu}$ taken over a hyperplane σ is independent on σ. Hence, choosing the hyperplane x^o=const. (i.e., the three-dimensional space, cf. p.91), then we obtain the above mentioned representation of the field momentum:

$$P^\nu = \frac{1}{c} \int d\sigma_\mu \ \Theta^{\mu\nu} = \frac{1}{c} \int d^3x \ \Theta^{o\nu} \qquad (6.53)$$

Since this integral does not depend on the choice of the hyperplane, P^ν must be vector.

While the canonical tensor $\Theta^{\mu\nu}$ describes, according to (6.49), the conservation of field momentum, it has some undesirable properties. To start with one notes that, because of the term containing the factor $\partial^\nu A_\lambda$, $\Theta^{\mu\nu}$ is not gauge invariant. But then $\Theta^{\mu\nu}$ may not be identified with the physical observables of energy- and momentum-density, because these ought to be gauge invariant.

In addition, the canonical tensor is not symmetric. Therefore, the momentum and angular momentum cannot be connected by a relation analogous to that which holds in classical mechanics. Indeed, the moments (angular momentum tensor)

$$M^{\lambda\mu\nu} = T^{\lambda\mu} x^\nu - T^{\lambda\nu} x^\mu \qquad (6.54)$$

formed in analogy to mechanics are conserved only if the divergenceless energy-momentum tensor $T^{\mu\nu}$ is symmetric. For if one forms $\partial_\lambda M^{\lambda\mu\nu} = 0$, then, with $\partial_\lambda T^{\mu\nu} = 0$ one is led to the symmetry condition

$$\partial_\lambda M^{\lambda\mu\nu} = T^{\nu\mu} - T^{\mu\nu} = 0 \qquad (6.55)$$

One can amend the canonical tensor which does not obey this condition to become the symmetric tensor $T^{\mu\nu}$ if one adds a suitable term $\partial_\lambda \phi^{\lambda\mu\nu}$, thus:

$$T^{\mu\nu} = \Theta^{\mu\nu} + \partial_\lambda \phi^{\lambda\mu\nu}, \quad \phi^{\lambda\mu\nu} = -\phi^{\mu\lambda\nu}, \qquad (6.56)$$

This is also gauge invariant. Demanding antisymmetry of the tensor

$\phi^{\lambda\mu\nu}$ in the first two indices ensures that $T^{\mu\nu}$ is divergenceless provided this holds for $\Theta^{\mu\nu}$. The total momentum of the field is not changed by the supplementary term. Nevertheless we emphasize here that this ad hoc and often used symmetrization procedure is not really needed in our case. Rather, one obtains the symmetric energy—momentum tensor of the electromagnetic field from Noether's theorem if one considers invariance against the homogeneous Lorentz transformations.

The infinitesimal Lorentz transformation

$$x'^{\mu} = x^{\mu} + \delta\epsilon^{\mu}{}_{\nu}x^{\nu}$$

leads to the following change of a four-vector A^{μ} (cf. equ.(3.7)):

$$\delta A^{\mu} = \delta\epsilon^{\mu}{}_{\nu}A^{\nu} .$$

According to Noether's theorem (6.47) we have

$$\partial_{\lambda}\left\{H^{\lambda\mu\nu} - \Theta^{\lambda\mu}x^{\nu}\right\}\delta\epsilon_{\mu\nu} = 0 , \qquad H^{\lambda\mu\nu} := \frac{\partial\mathcal{L}_F}{\partial A_{\mu,\lambda}}A^{\nu} , \qquad (6.57)$$

and taking into account the antisymmetry $\delta\epsilon_{\mu\nu} = -\delta\epsilon_{\nu\mu}$,

$$\partial_{\lambda}\left\{H^{\lambda\mu\nu} - H^{\lambda\nu\mu} - \Theta^{\lambda\mu}x^{\nu} + \Theta^{\lambda\nu}x^{\mu}\right\}\delta\epsilon_{\mu\nu} = 0 . \qquad (6.58)$$

Since the six independent parameters $\delta\epsilon_{\mu\nu}$ ($\mu < \nu$) are arbitrary, we obtain from above the six conservation laws

$$\partial_{\lambda}L^{\lambda\mu\nu} = 0$$
$$L^{\lambda\mu\nu} := \Theta^{\lambda\mu}x^{\nu} - \Theta^{\lambda\nu}x^{\mu} - H^{\lambda\mu\nu} + H^{\lambda\nu\mu} \qquad (6.59)$$

Sometimes this tensor is called the angular momentum tensor, but it does not have the property expressed by (6.54) and, besides, it is not gauge invariant.

However, if one actually calculates the derivatives in equation (6.58) then, with (6.49) it follows that

$$\partial_{\lambda}L^{\lambda\mu\nu} = \Theta^{\nu\mu} - \Theta^{\mu\nu} - \partial_{\lambda}H^{\lambda\mu\nu} + \partial_{\lambda}H^{\lambda\nu\mu} = 0$$

In this way one is led to the symmetric tensor

$$T^{\mu\nu} = \theta^{\mu\nu} + \partial_\lambda H^{\lambda\mu\nu} .$$

(6.60)

With the definition (6.57) of $H^{\lambda\mu\nu}$ and with (6.50) one gets

$$H^{\lambda\mu\nu} = -\frac{1}{4\pi} F^{\lambda\mu} A^\nu .$$

Since $H^{\lambda\mu\nu}$ is antisymmetric in the first two indices, the divergence of $\partial_\mu H^{\lambda\mu\nu}$ vanishes,

$$\partial_\mu \partial_\lambda H^{\lambda\mu\nu} = 0$$

and therefore, together with $\partial_\mu \theta^{\mu\nu} = 0$, we also have

$$\partial_\mu T^{\mu\nu} = 0$$

(6.61)

Thus, the symmetric and divergenceless tensor $T^{\mu\nu}$ satisfies the conditions (6.54) and (6.55) and apparently it is the correct energy-momentum tensor. The tensor $H^{\lambda\mu\nu}$ which comes here from Noether's theorem coincides with the tensor $\phi^{\lambda\mu\nu}$ which was required earlier for the purpose of symmetrization.

In charge-free space (which we now consider) we have $\partial_\lambda F^{\lambda\nu} = 0$, so that

$$\partial_\lambda H^{\lambda\mu\nu} = -\frac{1}{4\pi} F^{\lambda\mu} \partial_\lambda A^\nu ,$$

and therefore with (6.51)

$$T^{\mu\nu} = -\frac{1}{4\pi} F^{\mu\lambda} \left(\partial^\nu A_\lambda - \partial_\lambda A^\nu \right) + \frac{1}{16\pi} g^{\mu\nu} F_{\alpha\beta} F^{\alpha\beta}$$

This finally leads to the following expression for the symmetric energy-momentum tensor of the electromagnetic field:

$$T^{\mu\nu} = \frac{1}{4\pi} \left\{ g^{\mu\alpha} F_{\alpha\beta} F^{\beta\nu} + \frac{1}{4} g^{\mu\nu} F_{\alpha\beta} F^{\alpha\beta} \right\} .$$

(6.62)

Clearly, this tensor is also gauge invariant. In addition it has vanishing trace, $T^\mu_{\ \mu} = 0$. The differential conservation law for the angular momentum tensor

$$M^{\lambda\mu\nu} = T^{\lambda\mu} x^{\nu} - T^{\lambda\nu} x^{\mu}$$

formed with this $T^{\mu\nu}$ now reads

$$\partial_{\lambda} M^{\lambda\mu\nu} = 0 \quad .$$

Therefore, the total angular momentum of the field,

$$J^{\mu\nu} = \frac{1}{c} \int d\sigma_{\lambda} \, M^{\lambda\mu\nu} = \frac{1}{c} \int d^3 x \, M^{o\mu\nu} \tag{6.63}$$

is conserved. Because of the independence of the quantities $J^{\mu\nu}$ on the hypersurface σ, the total angular momentum of the field is a tensor.

The above obtained quantities become more clear to us if we express them in terms of the field strengths $\vec{E}$ and $\vec{B}$. With the energy density

$$W = \frac{1}{8\pi} \left(\vec{E}^2 + \vec{B}^2 \right) \tag{6.64a}$$

and with the momentum density

$$\vec{g} = \frac{1}{4\pi c} \left[\vec{E} \times \vec{B} \right] \tag{6.64b}$$

and with the Maxwell stress tensor

$$T^{mn} = \frac{1}{4\pi} \left\{ -E^m E^n - B^m B^n + \frac{1}{2} \delta_n^m \left(\vec{E}^2 + \vec{B}^2 \right) \right\} , \quad E^1 = E_x , \tag{6.64c}$$

one can summarize the energy-momentum tensor in the following matrix of components:

$$T^{\mu\nu} = \left(\begin{array}{c|c} W & c\vec{g} \\ \hline c\vec{g} & T^{mn} \end{array} \right) . \tag{6.65}$$

Taking $\nu = 0$, the conservation law (6.61) results in Poynting's law

$$\partial_\mu T^{\mu o} = \frac{1}{c}\left(\frac{\partial w}{\partial t} + \vec{\nabla}\cdot\vec{S}\right) = 0 \tag{6.66}$$

and taking $\nu = n$ we obtain the conservation law of the spatial field momentum,

$$\partial_\mu T^{\mu n} = \frac{\partial g^n}{\partial t} + \frac{\partial}{\partial x^m} T^{mn} = 0 \;. \tag{6.67}$$

We see that the above Poynting vector

$$\vec{S} = c^2\,\vec{g} \tag{6.68}$$

can be interpreted as the density of energy current, and the tensor T^{mn} can be interpreted as the density of momentum current. This follows directly from the formulation of the conservation laws in their integral form which can be read as equations of balance. We already explained this[1] in Section 6.9 (cf. equ.(6.45)).

According to equation (6.63) the spatial components of the antisymmetric tensor $J^{\mu\nu}$ are given by the formula

$$J^{mn} = \frac{1}{c}\int d^3x\left(T^{om}x^n - T^{on}x^m\right)$$

Since T^{om}/c stands for the components of momentum density, J^{32}, J^{13}, and J^{21} clearly represent the components of the field's total angular momentum $\vec{J}$. Using the expression (6.64b) for the momentum density, one obtains for $\vec{J}$ the expression

$$\vec{J} = \frac{1}{4\pi c}\int d^3x\;\vec{x}\times\left[\vec{E}\times\vec{B}\right] \;. \tag{6.69}$$

The mixed components of $J^{\mu\nu}$ give

$$J^{on} = -\,ct\;P^n + \frac{1}{c}\int d^3x\;w\,x^n$$

---

[1] One must observe that the density of the energy current is a three-dimensional vector and, correspondingly, the density of the momentum current, with the momentum being a vector, must be a three-dimensional tensor of rank two.

These quantities have a structure analogous to those in Section 5.6 (p.140), which, for a free particle, lead to the center-of-mass theorem. If we now correspondingly define the vector of the field's energy center of gravity by

$$\vec{X}_F = \frac{\int d^3x \, w \, \vec{x}}{\int d^3x \, w} \tag{6.70}$$

then from the temporal constance of J^{on} and from $P^\nu=(W/c,\vec{P})$ it follows that

$$\frac{d\vec{X}_F}{dt} = \frac{\vec{P}}{W/c^2} = \text{const} \tag{6.71}$$

This is the center-of-mass theorem for the free electromagnetic field. The velocity of the center of mass is constant and equals the field momentum divided by the total "field mass" W/c^2.

If charges are present, the inhomogeneous field equation (6.39) must be satisfied and the conservation law (6.61) is no longer valid. Since now $[\mathcal{L}_F]_{A_\mu}$ is different from zero, Noether's theorem leads to

$$\partial_\mu T^{\mu\nu} = -\frac{1}{c} F^{\nu\beta} j_\beta \tag{6.72}$$

This relation can be easily checked if one takes into consideration the field equations (6.39) and the property (6.25) of $F^{\mu\nu}$ when one calculates the divergence of $T^{\mu\nu}$. The divergence of $T^{\mu\nu}$ no longer vanishes but equals to the negative of the Lorentz force density

$$k^\nu := \frac{1}{c} F^{\nu\beta} j_\beta = \left(\frac{1}{c} \vec{j} \cdot \vec{E}, \, \varrho\vec{E} + \frac{1}{c} \left[\vec{j} \times \vec{B} \right] \right) \tag{6.73}$$

With spatial integration of (6.72) one obtains for $\nu = 0$

$$-\frac{\partial}{\partial t} \int d^3x \, w = \int d^3x \, \vec{\nabla} \cdot \vec{S} + \int d^3x \, \vec{j} \cdot \vec{E} \tag{6.74}$$

Thus, the previous energy balance (6.66) is amended by the integral over $\vec{j} \cdot \vec{E}$. This term characterizes the decrease of the field energy due to the work done, per unit time, by the field on the current $\vec{j}$ (on the charged particles). The magnetic field does not occur here, because on account of $\vec{B} \perp \vec{j}$ it does not do work. The Poynting vector in the first term allows for energy propagation in the form of electromagnetic radiation also in space devoid of matter. The second term indicates conversion of electromagnetic energy into mechanical or thermal energy due to work done on the charges present.

A corresponding balance holds for the change of the three-dimensional field momentum $\vec{P}$, which is determined by the Lorentz force (6.73) exerted by field on the charges and currents.

In summary, for a system consisting of the field and of charges (particles) the energy momentum tensor of the field is not conserved by itself, since we deal here with an open (i.e. not isolated) subsystem. Energy and momentum of the field decrease in the rate by which these quantities are transferred from the field to the charges and currents. When defining the conserved entities, we must now take into consideration the energy and momentum of the charges (particles). With Noether's theorem, a conserved energy-momentum tensor can be constructed only for the closed system of field and charges, and we now have

$$\partial_\mu T^{\mu\nu} = 0 \;, \qquad T^{\mu\nu} = T_F^{\mu\nu} + T_M^{\mu\nu} \tag{6.75}$$

Here the tensors $T_F^{\mu\nu}$ (field) and $T_M^{\mu\nu}$ (matter) of the subsystems are not separately divergenceless. Rather one obtains, since because of (6.72) and (6.73) the relation

$$\partial_\mu T_F^{\mu\nu} = -k^\nu \tag{6.76}$$

holds, the following equation:

$$\partial_\mu T^{\mu\nu} = -k^\nu + \partial_\mu T_M^{\mu\nu} = 0 \;,$$

that is,

$$\partial_\mu T_M^{\mu\nu} = k^\nu \tag{6.77}$$

The two subsystems mutually determine each other. On one hand, the fields are determined by the moving charges, on the other hand the motion of the charges is determined by the fields.

We saw that the conservation laws of energy, momentum, angular momentum, and center-of-mass motion for closed systems, which follow from the universal relativity principle, can be expressed in a particularly simple way in terms of the energy-momentum tensor. In this manner one finds the expressions for the observables of the system that must be identified with the conserved quantities. Knowledge of these quantities in terms of the fields is especially useful for the development of the corresponding quantum field theory, i.e., in the present case, quantum electrodynamics. This comment underscores the importance of the energy-momentum tensor, and consequently, of Noether's theorem.

Chapter 7

RELATIVISTIC HYDRODYNAMICS

For discussing the macroscopic behaviour of matter (solid bodies, fluids, gases) one may often begin with a material continuum which is described in terms of space and time dependent functions (field functions) such as the mass density $\rho(\vec{x},t)$, the velocity field $\vec{v}(\vec{x},t)$, the pressure field $p(\vec{x},t)$, and perhaps additional such entities. The particle index i which in point mechanics serves to distinguish the particles of mass m_i and velocity $\vec{v}_i$, is replaced in continuum mechanics by the continuous variable $\vec{x}$. In consequence, the motion of continuous media is described by partial differential equations. Since, in principle, the structure of matter is atomistic, one has here naturally a phenomenological description. Since in most applications the velocities which occur are small compared to the speed of light, one can rely on Newtonian mechanics. However, having in mind the universal validity of Einstein's relativity principle, it is of fundamental interest to formulate a relativistic mechanics of continua which, for $v \ll c$, goes over into the familiar classical theory.

Relativistic hydrodynamics, which is very suitable for clarification of concepts, nowadays has not only an academic interest. Rather, it plays an important role in the formulation of cosmological models and also in certain other problems of relativistic astrophysics, for example in cases when gases in strong gravitational fields achieve relativistic flow velocities. A particularly interesting and topical current application of relativistic hydrodynamics is in the domain of nuclear physics, specifically for models used to describe the phenomena which take place in the collision of relativistic heavy ions.

In the following we shall discuss the most important concepts and equations of relativistic hydrodynamics.[1] We restrict ourselves to the cases of pulverized (incoherent) matter and of an ideal fluid. It is best to begin with constructing the appropriate energy–momentum tensor. When doing so, we can refer back to our results concerning the energy–momentum tensor of the electromagnetic field, see Section 6.10. As we saw there, physical fields possess their own continuous momentum and energy distribution. In particular, momentum and energy density are represented by the components T^{on}/c and T^{oo}, respectively, of the energy–momentum tensor, while the components cT^{mo} and F^{mn} are interpreted as energy–current density and momentum–current density, respectively. For a closed system the divergence of the energy–momentum tensor vanishes, i.e. energy and momentum are conserved. From the conservation laws one can deduce the equations of motion, namely $\partial_\mu T_M^{\mu\nu} = k^\nu$. This procedure will prove handy for the derivation of the equations of motion in relativistic hydrodynamics.

7.1   The Nonrelativistic Equations

In nonrelativistic physics, hydrodynamics of ideal fluids and gases is based on the equation of continuity

$$\frac{\partial \varrho}{\partial t} + \vec{\nabla} \cdot (\varrho \vec{v}) = 0 \tag{7.1}$$

and the equation of motion (the Euler equation)

---

[1] Certain entities of continuum mechanics, such as pressure and density, also play a role as thermodynamical variables, which shows the close connection between continuum mechanics and thermodynamics. In view of the limited space available, we cannot elaborate here on the thermodynamic aspects which are important for realistic processes.

$$\varrho \left(\frac{\partial \vec{v}}{\partial t} + \vec{v} \cdot \vec{\nabla} \right) \vec{v} = - \vec{\nabla} p \qquad (7.2)$$

These equations must be completed with the equation of state,

$$p = p (\varrho) , \qquad (7.3)$$

which gives the connection between the density ρ and the pressure p.

The equation of continuity (7.1) expresses the conservation law of matter (of the mass). On the left-hand side of the Euler equation (7.2) we have the acceleration $d\vec{v}/dt$ multiplied by the mass density ρ. When calculating the differential quotient of the velocity field $\vec{v}(\vec{x},t)$ mentioned above, the coordinates must not be taken as if they were independent of t; instead, they are functions of t. So doing, one gets for each component of $\vec{v}$ the so-called substantial derivative (frequently referred to as differentiation following the particle)

$$\frac{d v_x}{d t} = \frac{\partial v_x}{\partial t} + (\vec{v} \cdot \vec{\nabla}) v_x , \qquad (7.4)$$

which is additively composed from the local temporal change $\partial v_x / \partial t$ and the contribution coming from the translation of the fluid volume element with the velocity $\vec{v}(\vec{x},t)$. According to Newton's basic law of mechanics, we must have on the right-hand side of equation (7.2) some force density. Ideal liquids are distinguished by permitting one to neglect processes with energy dissipation (internal friction (viscosity) and heat conduction). Thus, they do not show resistance to a shearing deformation, because their viscosity is negligible. Therefore, the (internal) forces exerted on the surface of a certain fluid volume element by neighbouring regions equal the surface integral taken over the pressure, which can be also expressed as a volume integral:

$$- \oint p d\vec{f} = - \int \vec{\nabla} p d V \qquad (7.5)$$

Consequently, the force density which occurs in the Euler equation, is equal to the negative gradient of the pressure. If there act external forces, the corresponding force density $\vec{k}$ must also be taken into consideration on the right-hand side of (7.2).

7.2 Conservation of Particle Number

As we saw in Section 5.2, the nonconservation of the rest mass m is an essential feature of relativistic theory. Therefore one cannot expect that, by introducing a four-vector for the mass current density, the continuity equation (7.1) would immediately become a covariant relation. Since in relativity theory it is not the rest mass but the energy which is conserved, one will have to take into account all occurring forms of energy, such as for example the elastic energy density of the fluid, if one wants to set up a correct balance equation. As we shall see for the case of an ideal fluid, this leads to an appropriately modified form of the nonrelativistic equation of continuity.

In contrast to the mass density, the particle number density $n_0(\vec{x},t)$ referred to the local rest system obeys an equation of continuity with the form $\partial_\mu(n_0 u^\mu) = 0$, where u^μ is the four-velocity of the fluid element under consideration. To see this, we start with the nonrelativistic equation. If no particles are created or annihilated, then the total number of particles is conserved, i.e. one has for the particle number density $n(\vec{x},t)$ the continuity equation

$$\frac{\partial n}{\partial t} + \vec{\nabla}(n\vec{v}) = 0. \tag{7.6}$$

One should note that $\vec{v}$ stands for the nonrelativistic velocity field and n is not a Lorentz scalar. If by a Lorentz transformation one goes over to a reference frame K^o in which the fluid element under consideration is at rest, then one must take into account the change of the three-dimensional volume element dV (see equ.(2.15) on p.38):

$$dV = \sqrt{1-\beta^2}\, dV_o \tag{7.7}$$

One can now introduce as a Lorentz scalar the particle number density in K^o, which we call the rest density n_o. Then one has

$$n \, dV = n_o \, dV_o \qquad (7.8)$$

since the particle number in a specified volume does not change under a Lorentz transformation. If one replaces in this equation dV by $\sqrt{1-\beta^2} \, dV_o$, it follows that

$$n = \gamma n_o \, , \qquad \gamma = \frac{1}{\sqrt{1-\beta^2}} \qquad (7.9)$$

One can use this relation in equation (7.6) and with the four-velocity $u^\mu = (\gamma c, \gamma \vec{v})$ (cf. p.103) one eventually obtains the equation

$$\frac{1}{c} \frac{\partial}{\partial t} (n_o \gamma c) + \vec{\nabla}(n_o \gamma \vec{v}) = \partial_\mu (n_o u^\mu) = 0 \qquad (7.10)$$

Since n_o is a scalar, $n_o u^\mu$ stands for a four-vector which has a vanishing divergence. Thus, the equation of continuity (7.6) becomes replaced in the relativistic case by equation (7.10).

If in high-energy processes new particles are created (that is to say particle-antiparticle pairs) then in the balance the particles must be taken into account with a positive contribution and the antiparticles with a negative one. This corresponds to the identification of n_o with the baryon number density, and hence to baryon number conservation. For an electron gas, the lepton number is conserved, i.e. one then must use in the continuity equation the lepton number density.

If one multiplies the particle number density n_o by the rest mass m of the particles of which the medium consists, then one obtains the mass density ρ_o measured in the local instantaneous rest system,

$$\rho_o = m \, n_o \qquad (7.11)$$

For the mass density one has the following equation of continuity in relativistic form

$$\partial_\mu (\rho_o u^\mu) = 0 \qquad (7.12)$$

only in the particularly simple case of a medium which consists of noninteracting particles with the same mass (relativistic dust). We wish now to study in detail this simple model of a material medium, after which we shall proceed to the more general case of an ideal fluid.

7.3   Incoherent Matter

Suppose the continuum we consider consists of a number of non-interacting particles with the same mass (relativistic dust). For such a system of incoherent material particles there are neither stresses nor an energy transport, so that the energy density in the instantaneous rest system arises from the rest mass alone. But the energy density is represented by the T^{oo} component of the energy-momentum tensor. Therefore, in the local rest system we have $T^{oo}=\rho_o c^2$, and with the four-velocity $u^\mu=c\delta^\mu_{\ o}$ we find

$$T^{\mu\nu} = \varrho_o c^2 \, \delta^\mu_o \, \delta^\nu_o \tag{7.13}$$

Since $T^{\mu\nu}$ must be a tensor and u^μ is a four-vector, we have to start in any arbitrary inertial system from the energy-momentum tensor

$$T^{\mu\nu} = \varrho_o \, u^\mu u^\nu \tag{7.14}$$

First of all we see that $T^{\mu\nu}$ is symmetric. It consists of the scalar field ρ_o and the vector field u^μ, which are the two entities that characterize a dustlike medium. In the local rest system the tensor leads back to (7.13). Furthermore, the density of the four-momentum is

$$\frac{1}{c^2} \, T^{\mu\nu} u_\mu = \varrho_o \, u^\nu \tag{7.15}$$

We note that this "projection" onto the direction of the four-velocity is equivalent to choosing the temporal component $T^{o\nu}$ to be evaluated in the instantaneous rest system, since the actual calculation of the summation in this system gives $T^{\mu\nu}u_\mu = T^{o\nu}u_o = T^{o\nu}c$.

The tensor (7.14) assumes a clearer form if its components

are expressed in terms of the three-dimensional velocity field $\vec{v}=(v_x,v_y,v_z)$. With $u^{\mu}=(\gamma c,\gamma \vec{v})$ one obtains for $T^{\mu\nu}$ the following matrix

$$T^{\mu\nu} = \rho_0 \gamma^2 \begin{pmatrix} c^2 & c\,v_x & c\,v_y & c\,v_z \\ c\,v_x & v_x^2 & v_x v_y & v_x v_z \\ c\,v_y & v_y v_x & v_y^2 & v_y v_z \\ c\,v_z & v_z v_x & v_z v_y & v_z^2 \end{pmatrix} \qquad (7.16)$$

For a closed system (no external forces) the energy-momentum tensor is divergenceless, i.e.

$$\partial_{\mu} T^{\mu\nu} = 0 \qquad (7.17)$$

From this, two important equations follow, viz. the already familiar equation of continuity (7.12) and the relativistic equation of motion. To see this, one performs the differentiation in (7.17), multiplies the result[1]

$$(\rho_0\, u^{\mu})_{,\mu}\, u^{\nu} + \rho_0\, u^{\mu} u^{\nu}_{,\mu} = 0 \qquad (7.18)$$

by u_{ν} and sums over ν. (This is a contraction with u_{ν}.) In this way the expression (7.18) is "projected" onto the direction of the four-velocity. Since $u^{\nu} u_{\nu} = c^2$ and therefore $u^{\nu}{}_{,\mu} u_{\nu} = 0$, equation (7.18) leads to

$$(\rho_0\, u^{\mu})_{,\mu} = 0 \qquad (7.19)$$

This is precisely the continuity equation (7.12) for the mass current density in our simple model. Taking it into account in equation (7.18), it follows that

$$\rho_0\, u^{\nu}_{,\mu}\, u^{\mu} = 0 \qquad (7.20)$$

To see the meaning of this equation, one goes to the nonrelativistic limit by neglecting v/c against 1, i.e. setting $\gamma=1$. With $u^0 \to c$, $u^n \to v^n$, the spatial components of (7.20) give

---

[1] Here again the following shorthand notation is used: $\partial_{\mu}\psi = \psi_{,\mu}$.

$$\varrho_0 \, v_{,0}^n \, c + \varrho_0 \, v_{,m}^n \, v^m = 0$$

or, in the notation using three-dimensional vectors,

$$\varrho \left(\frac{\partial}{\partial t} + \vec{v} \cdot \vec{\nabla} \right) \vec{v} = 0 \tag{7.21}$$

One must note that the mass density ρ_0 is defined in the local rest system, while in the nonrelativistic equation (7.21) the mass density ρ is referred to the unit volume in the laboratory frame. Therefore one must make the substitution $\rho_0 \to \rho \sqrt{1-\beta^2}$ when performing the limit. In the present case it suffices to take $\rho_0 \to \rho$.

In summary, the equations (7.20) are the relativistic generalization of the Euler equations for dustlike matter. Taking into account that the proper time derivative of the velocity field $u^\nu(x)$ is given by

$$\frac{du^\nu}{d\tau} = \frac{\partial u^\nu}{\partial x^\mu} \frac{dx^\mu}{d\tau} = u_{,\mu}^\nu \, u^\mu \tag{7.22}$$

the equations (7.20) can also be written in the form

$$\varrho_0 \frac{du^\nu}{d\tau} = 0$$

This form of the equations corresponds precisely to the equations of motion of a system of free particles as familiar from point mechanics.

7.4 The Ideal Fluid

The method used in the preceding section for the derivation of the equations characterizing relativistic hydrodynamics in the special case of incoherent matter may be directly applied to more general cases. Since in the model of an ideal (or perfect) fluid to which we now turn no frictional forces are acting, apart from the energy density we must take into account in the relevant energy-momentum tensor only the contribution of pressure. The forces caused by pressure act on the surface elements of the neighboring volume elements of the fluid. According to Section 6.10, the stress tensor T_{ik} $(i,k=1,2,3)$

can be interpreted as the density of the momentum current (cf. p.181). In particular, the balance expressed in equation (6.67) says that the momentum flux through the surface element $\vec{df}$ of a volume equals the force on that element. Thus, $T_{ik}df_k$ is the i-component of the force acting on a surface element. As before, we choose a reference frame in which the considered volume element of the flowing liquid is at rest. In this local rest system Pascal's law holds, according to which the pressure exerted by a given portion of fluid is the same in all directions and is perpendicular to the area on which it acts. Accordingly, the i-component of the force is also pdf_i. Hence, in the local rest system

$$T_{ik}df_k = pdf_i$$

so that

$$T_{ik} = p\,\delta_{ik}$$

The momentum density components $g^k = T^{ok}/c$ vanish in the rest system of the volume element considered, and the component T^{oo} represents the energy density. Let us denote the total energy density in the rest system by ε. The corresponding equivalent mass density is then $\mu \equiv \varepsilon/c^2$. Note that in the general case of an ideal fluid the quantity ε measured in the local rest system consists not only of the contribution $\rho_o c^2$ which stems from the mass density proper ρ_o, but it contains also the inner energy density e which arises from the microscopic particle motions and the interaction of the particles. Thus, in the local rest system the energy-momentum tensor has the following form:

$$T^{\mu\nu} = \begin{pmatrix} \mu c^2 & 0 & 0 & 0 \\ 0 & p & 0 & 0 \\ 0 & 0 & p & 0 \\ 0 & 0 & 0 & p \end{pmatrix} \qquad (7.23)$$

From this one can then find easily its form in an arbitrary inertial system if one goes with a Lorentz transformation to a reference frame that is at rest in the laboratory. A quicker procedure is furnished by the following consideration. To form $T^{\mu\nu}$ one can use only the two tensors $u^{\mu}u^{\nu}$ and $g^{\mu\nu}$. In the local rest system the tensor we are looking for must have the form (7.23). Using the scalars c^2 and p, the tensor obeying these conditions can be written as a linear combination of $u^{\mu}u^{\nu}$ and $g^{\mu\nu}$:

$$T^{\mu\nu} = \left(\mu + \frac{p}{c^2}\right) u^{\mu}u^{\nu} - pg^{\mu\nu} \tag{7.24}$$

This expression represents the energy-momentum tensor of an ideal fluid which we are looking for. One easily convinces oneself that in the rest system ($u^{\mu} = (c,\vec{0})$) this tensor goes over into the expression (7.23).

The tensor (7.24) is symmetric. Contracting $T^{\mu\nu}$ with u_{μ} leads to the four-momentum density:

$$\frac{1}{c^2} T^{\mu\nu}u_{\mu} = \mu\, u^{\nu} \tag{7.25}$$

Analogously to the case of the model of incoherent matter which was discussed in detail in Section 7.3, here we can again find the equation of continuity and the equation of motion by utilizing the conservation law (7.17) valid for closed systems. In contrast to the simpler case of Section 7.3, by forming the divergence of (7.24) and equating this with zero, here one must consider also the derivatives of the pressure $p(x)$. After contracting with u_{ν} and dividing by c^2 (note again that $u^{\nu}u_{\nu} = c^2$) the result of calculating the divergence,

$$(\mu\, u^{\mu})_{,\mu} u^{\nu} + \mu\, u^{\mu}u^{\nu}_{,\mu} + \left(\frac{p}{c^2}\right)u^{\mu}_{,\mu}u^{\nu} + \left(\frac{p}{c^2}\right)u^{\mu}u^{\nu}_{,\mu} + c^{-2} p_{,\mu}u^{\mu}u^{\nu} - p_{,\mu}g^{\mu\nu} = 0 \tag{7.26}$$

leads to

$$(\mu\, u^{\mu})_{,\mu} + \frac{p}{c^2}\, u^{\mu}_{,\mu} = 0 \tag{7.27}$$

This is obviously the generalized continuity equation for the mass density μ equivalent to ε. We see that neither the mass density proper ρ_{o} nor μ satisfies a conservation law of the form (7.19).

In contrast to the case of incoherent matter, in equation (7.27) we have also the contribution of the pressure. We can expect this if we realize that, in relativity, it is not the rest mass but the energy which is conserved and the pressure of a fluid contributes to its energy content.

Taking into account the result (7.27), the equation (7.26) can be written in the following simpler form

$$\left(\mu + \frac{p}{c^2} \right) u^{\nu}_{,\mu} u^{\mu} = \left(g^{\mu\nu} - \frac{u^{\mu} u^{\nu}}{c^2} \right) p_{,\mu} \tag{7.28}$$

This relation represents the relativistic equation of motion of an ideal fluid. For vanishing pressure (i.e. $p=0, p_{,\mu}=0$) it assumes the form (7.20). Using (7.22) and $u^{\nu} = dx^{\nu}/d\tau$, (7.28) may also be written as

$$\left(\mu + \frac{p}{c^2} \right) \frac{d^2 x^{\nu}}{d\tau^2} = \left(g^{\mu\nu} - \frac{u^{\mu} u^{\nu}}{c^2} \right) p_{,\mu} \tag{7.29}$$

which corresponds to the familiar form of an equation of motion. We see from this that the volume elements with inertial mass $\mu + p/c^2$ are prevented from free motion ($d^2 x^{\nu}/d\tau^2 = 0$) precisely by the pressure gradient $p_{,\mu}$.

As expected, for low velocities and small pressures the equations (7.27) and (7.28) go over into the classical equations (7.1) and (7.2), respectively. The transition to the nonrelativistic limiting case is performed, as in Section 7.3, with the substitutions $u^0 \to c$, $u^n \to v^n$ ($\gamma = 1$). For small pressures, p/c^2 may be neglected against μ. In addition, in the nonrelativistic limit the contribution $\rho_0 c^2$ of the rest energy density dominates in $\varepsilon = \mu c^2$, so that one is permitted to replace μ by ρ_0. When doing so, the contribution e/c^2 of the inner energy density (which is much smaller than ρ_0) is neglected. Furthermore, when going to the laboratory frame, one makes the replacement $\rho_0 \to \rho$. In this manner, one obtains from (7.27) the nonrelativistic equation of continuity (7.1). Regarding the equation of motion (7.28), one first observes that in the nonrelativistic limit the temporal component becomes zero on both sides. The spatial components simplify to

$$\rho_o \left(\frac{\partial v^n}{\partial t} + \frac{\partial v^n}{\partial x^m} v^m \right) = - \delta_{mn} \frac{\partial p}{\partial x^m}$$

Summed up in vector notation and substituting ρ for ρ_o, this is exactly the Euler equation (7.2) of an ideal fluid which is unaffected by external sources.

We saw that the basic equations of relativistic hydrodynamics can be derived from the conservation law $\partial_\mu T^{\mu\nu} = 0$ for the energy-momentum tensor. But one could reverse the argument, i.e. conclude $\partial_\mu T^{\mu\nu} = 0$ from the valid basic equations. In the more general case of nonideal fluids, where dissipative processes occur, the basic equations and hence the energy-momentum tensor must be generalized by appropriate additional terms. However, even for such a more complicated energy-momentum tensor one may start (without knowing better its actual form) from the property of its symmetry and the fact that for a closed system its divergence vanishes.

Chapter 8

LIMITS OF SPECIAL RELATIVITY

As already pointed out in the Introduction, the essential statement of special relativity theory is that of the covariance (form
invariance) of laws of nature against the change of inertial frames
according to the Lorentz group. Due to this symmetry the physical
laws which are formulated within the framework given by Einstein's
principle of relativity, are determined to a larger extent as relations
between tensor quantities. This should have become clear in course
of the preceding discussions of the laws of relativistic mechanics,
electrodynamics, and hydrodynamics.

The reason for the attribute "special" and the attendant limitation has often been misunderstood. The limitation says solely that
the theory of special relativity is valid only for situations where
one can ignore gravitational effects. The occasionally announced statement that special relativity applies only to motions with no accelerations (so-called inertial motions), is simply false. Indeed we saw
in Section 5.1 that the accelerated motion of a particle which occurs
under the influence of a four-force K^ν, can be calculated with the
relativistic equation of motion (5.4). As a specific example we refer
in particular to the equation of motion (6.14), according to which
the acceleration of a charge in an external field is caused by the
Lorentz force.

On the other hand, when applying the theory of special relativity, one must take notice of the fact that statements obtained with

the metric tensor (4.1) of the Minkowski space are valid only in unaccelerated frames of reference, i.e., they are restricted to inertial
systems. This limitation to inertial systems, related to the metric
(4.1), is often overlooked. In an accelerated frame of reference there
are inertial forces. Hence, such reference systems are not equivalent
to inertial systems. The corresponding metric tensor has no longer the
simple form (4.1) but depends explicitly on the coordinates. The line
element is then expressed in terms of curvilinear coordinates.

The essential limitation of special relativity theory that
we mentioned above consists in the fact that it is incompatible with
Newtonian gravitation theory, and indeed the latter cannot be simply
generalized to a consistent theory of gravitation in Minkowski space.
The conflict of Newtonian gravitation theory with special relativity
theory manifests itself in various ways. According to the Newtonian
law the attractive force between two bodies of mass m and M at a distance $r = (x^2+y^2+z^2)^{1/2}$ is

$$\vec{F} = -G\,\frac{mM}{r^3}\,\vec{r}\ .$$

$$(8.1)$$

Here $G = 6.67 \times 10^{-11} \mathrm{Nm^2 kg^{-2}}$ denotes the gravitational constant respon-
sible for this interaction. Relation (8.1) is covariant against spatial
rotations which clearly leave the distance r invariant. This applies
also to the corresponding Poisson equation

$$\vec{\nabla}^2\phi(\vec{x}) = 4\pi G\varrho(\vec{x})\ ,$$

$$(8.2)$$

where ϕ is the gravitational potential of a mass distribution with
density ρ. From this equation one obtains for a pointlike mass M
located at the coordinate origin (and with the appropriate boundary
conditions) the potential $\phi = -GM/r$, and forming the negative gradient
of $m\phi$ gives exactly the force (8.1).

However, in the theory of special relativity the symmetry group
consists of the Lorentz transformations which leave the square of
separation $s^2 = c^2 t^2 - x^2 - y^2 - z^2$ invariant. But against the Lorentz group
the equations (8.1) and (8.2) are not covariant. In physical terms

this means that in the Newtonian theory the velocity of light is not an invariant entity, neither has it the meaning of a limiting speed. Gravitation here acts instantaneously , i.e. perturbations of the gravitational field propagate with an infinite velocity. In Newton's formulation gravitation is described by an action-at-a-distance theory.

How can one avoid this conflict? Since special relativity theory proved itself excellently in all experimental tests (see Appendix B), one would first try to generalize Newton's law so that it becomes compatible with the theory of special relativity. A plausible generalization of the static equation (8.2) leads to the wave equation for the scalar gravitational potential ϕ :

$$\frac{1}{c^2} \frac{\partial^2 \phi}{\partial t^2} - \vec{\nabla}^2 \phi \equiv \Box \phi = -4\pi G \rho \ . \tag{8.3}$$

Indeed, this equation describes the propagation of perturbations of the gravitational field with light velocity, but it contradicts experience because this scalar theory would not result in the deflection of light in the gravitational field of the Sun. This is so because, if ϕ is a scalar field, then, according to (8.3), the source ρ of this field must be also a scalar. On the other hand, because of the equivalence of mass and energy one expects that any energy density is source of a gravitational field, just as a mass density. Now, the energy density is represented by the T^{oo} component of the energy-momentum tensor. Thus, if one wants to take into consideration the mass equivalent to the energy density of the electromagnetic radiation field, then on the right-hand side of equation (8.3) one should put the scalar quantity $g_{\mu\nu}T^{\mu\nu}/c^2$ in the place of the mass density. But one easily convinces oneself that this expression, i.e. the trace of the energy-momentum tensor of the radiation field (6.62) gives zero. Since the photons have no rest mass, it is impossible to have another coupling of the gravitational field to the radiation field, which means that in the scalar theory a light ray cannot be deflected by the gravitational field.

Obviously, for the gravitational field additional degrees of freedom are needed. The assumption of a vector field would lead us back to electrodynamics. The tensor of next higher rank is that with two indices. Thus it is only natural to generalize the wave equation (8.3) to the tensor equation

$$\Box \phi_{\mu\nu} = - \frac{16\pi G}{c^4} T_{\mu\nu} \qquad (8.4)$$

which is covariant against Lorentz transformations. One should observe, however, that equation (8.4) is valid in this simple form only if a special gauge ($\partial_\mu \phi^{\mu\nu} = 0$) is used. This is analogous to the Lorentz gauge of electrodynamics. But now, how should one interpret the tensor $\phi_{\mu\nu}$? Clearly, with equation (8.4) one gives up the simple structure of Newtonian gravitation theory. Even though in this tensor theory light deflection can be correctly described, for the perihelion precession it predicts a value which does not agree with the observations on the planet Mercury. Moreover it turns out that this theory of gravitation, formulated in Minkowski space, is not selfconsistent, since on one hand one requires $\partial_\nu T^{\mu\nu} = 0$ and must exclude it, on the other hand, due to the equation of motion.[1]

Apart from these discrepancies, it is not difficult to see that the tensor equation (8.4) in the best case can give only an approximation to a complete and consistent theory of gravitation. For, while equation (8.4) has on the left-hand side the linear differential operator of the wave equation, the gravitation theory which is supposed to generalize the Newton law, must be nonlinear. Again, we recall here the equivalence of mass and energy (Section 5.2). Due to this equivalence, the gravitational field produced by some mass distribution possesses a certain energy density which is equivalent to a mass density. In turn, this mass density represents a source of gravitation and contributes to the gravitational field. That is to say, the field itself can serve as part of its own source. Taking into account this direct self-interaction or feedback effect of the gravitational field, one realizes that relativistic gravitation theory is a nonlinear

---

[1] More about this one finds, for example, in Ch. W. Misner, K.S. Thorne, and J.A. Wheeler: Graviation (W.H. Freeman, San Francisco, 1973), p. 186.

theory, and correspondingly the field equations must be more complex than the simple linear relation (8.4). A direct self-interaction does not occur in a linear theory. Thus the Maxwell equations of electrodynamics are linear. To look at this example more closely, note that the electrostatic field of a charge contains a certain amount of electric field energy. But this does not produce an additional charge which would result in a direct self-interaction of the field and hence would render the theory nonlinear.

In summary, it is not possible to formulate a consistent theory of gravitation, compatible with observational facts, if one stays within the special theory of relativity, i.e. if one thinks of a theory in Minkowski space. A stepwise elimination of the named discrepancies of the tensorial theory mentioned above would eventually lead to the general theory of relativity given by Einstein, in which gravitation is expressed through the curvature of the space-time continuum. But then the geometry is no more pseudo-Euclidean as in Minkowski space, rather, because of the presence of curvature, the geometry is that of a Riemannian space, with a metric determined by the coordinate dependent tensor $g_{\mu\nu}(x)$ (cf. Section 3.4). Thus, when one attempts to relativistically generalize Newton's theory of gravitation, the framework of special relativity turns out to be too narrow. Therefore, equation (8.4) can be considered only as a linearized approximation of Einstein's gravitation theory. In this picture, the field functions $\phi_{\mu\nu}$ describe the deviations of the metric of curved space-time from that of flat Minkowski space; for weak gravitational fields these deviations may be assumed to be small enough such as to provide linearized field equations in this approximation.

The limitations of special relativity theory can also be recognized, in particular, from the fact that gravitational phenomena are not compatible with the concept of a (global) inertial frame for large regions of the space-time continuum. We recall that this concept, fundamental as it is for special relativity, is based on the law of inertia and is defined in the following way: In an inertial frame a body, not acted upon by forces, moves with constant velocity along a straight line. But how can one realize a force free state? While

it is possible to shield a material body against electromagnetic interactions and isolate it from other influences, this cannot be done, not
even in principle, for gravitation which is always an attractive force.
Hence one should imagine an inertial frame as far as possible from all
gravitating matter. But this makes the concept of an inertial frame
problematic, since in our surrounding universe matter is so abundantly
present that it would be difficult to maintain a sufficiently large
distance from all matter. In addition, there is a circular argument in
the law of inertia used for such a definition of an inertial frame.
According to this argument, a mass moves without acceleration if it is
sufficiently removed from other bodies; in turn, the fact that it
is sufficiently removed is recognized exactly by seeing that it moves
without acceleration. Because of all this, the inertial frame used
in the special theory of relativity is a much idealized concept. After
all, the experiments whose outcome we describe and interpret in terms
of physical laws, are performed on Earth, i.e. in the vicinity of
masses. And in the presence of an inhomogeneous gravitational field
an inertial frame can be realized only approximately.

To solve the listed difficulties when one formulates a relativistic theory of gravitation, necessitates a thorough rethinking, as
Einstein demonstrated in his general theory of relativity. If gravitation cannot be "switched off", one should take as a starting point
that gravitation and inertia are one and the same thing. Their effects
are, after all, always the same: gravitational forces, as well as
inertial forces, are proportional to the mass of the relevant body.
Experience shows that in a given gravitational field, all pointlike
particles fall with the same acceleration. This is possible only if
the inertial and gravitational mass of a body, introduced in mechanics
at first as different quantities, are equal to one another. Equality
of inertial and gravitational mass is an experimental fact of great
accuracy (Eötvös experiment), which has no explanation in classical
mechanics.

This equality cannot be a coincidence, rather, according to
Einstein it points to the essentially identical nature of gravitation
and inertia which must be incorporated into the framework of a

relativistic theory of gravitation. This consideration leads to Einstein's heuristic basic idea, the equivalence principle. According to this principle, no observer can himself distinguish the effects of a uniform constant acceleration on his instruments of observation and measurement from the effects of a homogeneous gravitational field exerted on the observer at rest. Accordingly, the action of gravitational fields is equivalent to an acceleration of the frame of reference.[1] Thus, if a (nonrotating) laboratory is in free fall in the gravitational field with the acceleration g, gravitation is cancelled everywhere where the acceleration is constant and equals g. The observer in free fall is in an inertial system. This is the cause of weightlessness in spacecrafts. But the acceleration caused by inhomogeneous gravitational fields is not spatially constant. Therefore, inhomogeneity of the gravitational field leads to relative acceleration of neighbouring bodies. This implies that gravitational effects can be transformed away by going into an accelerated frame of reference only locally, i.e. in regions of space in which the field may be taken homogeneous. An inhomogeneous gravitational field can be compensated only if one refers in each world point to another group of inertial systems. But in this way, a complicated metric connection between world points replaces the gravitational field. Whereas in the neighbourhood of a given world point the line element has the simple pseudo-Euclidean form (2.13) in the inertial system that was chosen at that point, nevertheless the inertial system in a neighbouring point is, in general, accelerated. Here the line element will have the general form (3.17), because for an accelerated reference system it is expressed by curvilinear coordinates. In consequence, the geometry of the space-time continuum will no longer be that of a (pseudo-) Euclidean space, but rather it corresponds to the geometry of a curved (Riemannian) space. From a mathematical point of view, the above-mentioned local inertial systems represent the plane tangential spaces of the Riemannian space at each particular

---

[1] A critical discussion of various formulations of the equivalence principle can be found in H.C. Ohanian, Am J. Phys. 45. 903 (1977).

point. Application of the Lorentz transformations is restricted to these spaces.

The gravitational field is described by the ten components of the metric tensor $g_{\mu\nu}(x)$; these functions obey the Einstein field equations. These equations are the already mentioned nonlinear generalization of the tensor equation (8.4), where again the energy-momentum tensor $T_{\mu\nu}$ appears as an inhomogeneity. However, the field functions $g_{\mu\nu}(x)$ are not fully determined by the field equations. This indeterminateness stems from the possibility of choosing coordinates arbitrarily; these are not prescribed by the field equations which are covariant against general transformations in the Riemannian space.

In summary, the gravitational field $g_{\mu\nu}(x)$, and consequently the metric of the Riemannian space, is generated by the mass and energy distribution present. The motion of masses in the gravitational field then occurs along geodesic lines of the space whose curvature is determined by the mass and energy distribution.

Thus, gravitation becomes a property of space-time and hence it is geometrized. Therefore, Einstein's gravitation theory can be also properly called "geometrodynamics". The geometric structure of the space-time continuum is determined by the matter present. In turn, this structure determines the motion of matter and prescribes the behaviour of meter rods and clocks. In the general theory of relativity inertia, metric, and gravitation are united. Space and time have no independent and hence no absolute existence, but are most closely related to objects that fill the space.

Appendix A. PROBLEMS

1. Convince yourself that nonlinear transformations of t and x lead to accelerations. (p.22) [1]

2. Show that the Lorentz transformations form a group. (p.26)

3. Using the equations (2.6), show that $s^2 = c^2t^2 - x^2$ is an invariant quantity. (p.28)

4. Assuming that the photon (in the optical domain) has a nonvanishing mass $m_\gamma \leq 10^{-59}$ g, what accuracy would be needed in the measurement of the light velocity c, so as to assert that it is different from the limiting velocity σ ? (p.29)

5. Derive the Lorentz transformation from the invariance principle formulated on p.36. Exclude scale transformations.

6. Can the writing speed of an oscilloscope (i.e. the speed of the produced light spot) be larger than the light velocity? If so, discuss the relevant conditions. (p.46)

7. Referring to the example discussed on p.47f, calculate the angle in the system K' between the table top and the x' axis.

8. Two rockets before start are at a distance L_o behind one another and are connected by a thin, unstretchable rope of proper length L_o, so that it is tightly stressed. At t = 0 both rockets are started with exactly the same constant acceleration relative to system K. At $t = t_1$ the acceleration terminates and both rockets move on with the same constant velocity (measured in K). Why will the rope break?

---

[1] Page numbers in these Problems refer to the connection of the problem with the relevant part of the main text.

9. In the example (discussed on p.49) where a man brings with high velocity a pole into a garage, three events are of importance. The first occurs when the front end of the pole reaches the line of the garage door; the second, when it hits the rear wall of the garage; and the third, when the rear end of the pole comes to the garage door. Events two and three occur simultaneously in K, which is the rest system of the garage. In K one can close the door at the instant of the third event. Show that in the man's system K' the third event (closing of the door) occurs later than the second and hence the garage door can be closed behind the pole also in K'. It will be advantageous to choose coordinates in such a way that for the first event $t = t' = 0$.

10. Prove the following statements concerning four-vectors:
If $A^\mu B_\mu = 0$ and A^μ is timelike, then B_μ must be spacelike.
If A^μ and B^μ are timelike vectors pointing into the future $(A^0 > 0, B^0 > 0)$, then their sum is also a timelike vector pointing into the future.
If two lightlike vectors are orthogonal to each other, then they are also parallel. (p.84f, Section 4.1)

11. Prove the statement made on p.163: If two four-vectors have a particular one of their components equal in all inertial systems, then they are identical. (p.88f)

12. Compute the curve in Fig.2.1, p.16. (p.115)

13. What is the minimal energy (in MeV) of a γ ray that can decompose a deuteron into its constituents?
$(m_p = 1.6726 \times 10^{-24} g; \quad m_n = 1.6748 \times 10^{-24} g; \quad m_d = 3.3435 \times 10^{-24} g).$ (p.116)

14. Compute the annual mass loss of the Sun. Based alone on this, how long would the Sun shine? Assume that only 50% of the total solar mass is available for energy production.
(Solar constant: $K = 2$ cal/cm^{2}min; distance of Earth to Sun: 150×10^6 km; mass of the Sun: 1.94×10^{33} g). (p.119)

15. Let K be a reference frame in which a system of particles has the total four-momentum $(E/c, \vec{P})$. Determine the special Lorentz transformation taking into the center-of-mass system K^*. Show that, relative to K, the velocity of the center-of-mass is $\vec{V}^* = c^2 \vec{P}/E$. (p.126)

16. Convince yourself that because of four−momentum conservation, an isolated free electron cannot emit or absorb a photon. (p.126f)

17. A particle of rest mass m_1 and velocity $\vec{v}_1$ is absorbed by a par−ticle of mass m_2 at rest. Determine the rest mass m and velocity $\vec{v}$ of the resulting system. (p.126f)

18. Consider inverse Compton scattering. A charged particle of rest mass m and lab−frame energy $E \gg m$ collides with a photon of fre−quency ν ($h\nu \ll m$). What is the maximal energy which the particle can transfer to the photon? Apply your result to calculate the energy transfer for the case when a cosmic ray proton ($m_p = 938$ MeV, $E = 10^{14}$ MeV) hits a photon from the cosmic background radiation of 3° K. (p.133)

19. Compute the field of a charged plane plate condenser of infinite extension which moves with velocity v parallel to its plates. Use the transformation formulae (6.20). (p.152)

20. Compute the field $\vec{E}'(x')$ of a moving point charge, given on page 154.

21. The Lagrangian density

$$\mathcal{L} = -\frac{1}{16\pi} F_{\alpha\beta} F^{\alpha\beta} + \frac{m^2}{8\pi} A_\alpha A^\alpha - \frac{1}{c} j_\alpha A^\alpha$$

differs from the one given on p.167 (equ.(6.37)) by the mass term $m^2 A_\mu A^\mu$.

a) Derive from the Lagrangian the corresponding field equations (the Proca equation). (p.167f)

b) Show that now the Lorentz condition follows from the equation of continuity $\partial_\mu j^\mu = 0$, and set up the wave equation for the potentials. What condition must satisfy the wave number vector of a plane wave which should be a solution of the corresponding homogeneous wave equation? (p.161)

c) What is the potential one obtains for a charge at rest at the origin?

Appendix B. THE EXPERIMENTAL TESTS OF SPECIAL RELATIVITY

In the following table[1] we list, with brief explanations, a number of more recent experiments which served to test the predictions of special relativity theory. In each experiment the effect of interest was measured at two velocities β_1 and β_2. The resolution R equals the experimental uncertainty devided by the nominal value of the effect. Some of these experiments have been quoted already in the text.

[1] T.S. Jaseja,et al., Phys.Rev.133, A 1221 (1964).

[2] R.F.C.Vessot and M.W.Levine,Metrologia 6, 116 (1970).

[3] R.V. Pound and G.A.Rebka, Jr.,Phys.Rev.Lett. 4, 274 (1960).

[4] H.Hay,J.Schiffer,T.Cranshaw,and P.Egelstaff,P.R.Lett.4,165 (1960).

[5] J.C.Hafele and Richard E.Keating; Science 177, 166 (1972).

[6] H.I.Mandelberg and L.Witten, J.Opt.Soc.Am. 52, 529 (1962).

[7] K.Brecher, P.R.Lett. 39, 1051, 1236 (E) (1977).

[8] D.J.Grove and J.G.Fox, Phys.Rev. 90, 378 (1953)

[9] V.P.Zrelow, A.A.Tyapkin, and P.S.Farago, Zh.Eksp.Teor.Fiz. 34, 555 (1958) (Sov.Phys. JETP 7, 384 (1958).

[10] D.S.Ayres, et al., Phys.Rev.D 3, 1051 (1971).

[11] T.Alväger, et al., Phys.Lett. 12, 260 (1964).

[12] Z.G.T. Guiragossián, et al., P.R.Lett. 34, 335 (1975).

[13] J.Bailey, et al., Nature 268, 301 (1977).

[14] J.Bailey, et al., Phys.Lett. 68 B, 191 (1977).

[15] J.Bailey, et al., Nuovo Cimento A9, 369 (1972).

[16] R.Van Dyck, P.Schwinberg, and H.Dehmelt, P.R.Lett. 38, 310 (1977).

[17] J.Wesley and A.Rich., Phys.Rev. A 4, 1341 (1971).

---

[1] With small alterations only, this table was taken from the paper by D. Newman, G.W. Ford, A. Rich, and E. Sweetman, Phys. Rev. Letters 40, 1355(1978). I thank here Prof. G.W. Ford and the publishers of The Physical Review Letters for their kind courtesy.

Ref.	Effect measured	Method	β_1	β_2	R
1	ether drift	Michelson-Morley interference	0	10^{-4}	10^{-3}
2	quadratic Doppler effect	temperature dependence of a H-maser	9×10^{-6}	10^{-5}	3×10^{-2}
3	quadratic Doppler effect	Mössbauer effect	2×10^{-4}	4×10^{-4}	10^{-1}
4	transversal Doppler effect	Mössbauer effect	0	7×10^{-7}	4×10^{-2}
5	time dilation	atomic clocks	10^{-6}	2×10^{-6}	3×10^{-2}
6	quadratic Doppler effect	spectral lines of moving atoms	0	7×10^{-3}	5×10^{-2}
7	velocity of ligth	timing of pulses from binary star	10^{-3}	-10^{-3}	2×10^{-9}
8	relativistic mass	moving protons	0	0.7	6×10^{-4}
9	relativistic mass	moving protons	0	0.81	10^{-3}
10	pion lifetime	decaying pions in a beam	0	0.92	4×10^{-4}
11	velocity of ligth	$\pi^{\circ} \to 2\,\gamma$	0	0.99975	1.3×10^{-4}
12	electrons of high energy	comparison with c	$1\,(\gamma)$	$1 - 5 \times 10^{-10}$	2×10^{-7}
13	muon lifetime	storage ring	0	0.9994	10^{-3}
14 15	g-factor of muon	precession in storage ring	0.38	0.9994	2.7×10^{-7}
16 17	g-factor of electron	precession in electromagnetic trap	5×10^{-5}	0.57	3.5×10^{-9}

Index

THE STENCILLED HOME

THE STENCILLED HOME

13 themed room styles

HELEN MORRIS
OF THE STENCIL LIBRARY

THE OVERLOOK PRESS

WOODSTOCK • NEW YORK

Acknowledgements

I extend my thanks to my talented friends who made my ideas and sketches real: to Geoff Garrett of Newcastle for the joinery and to Stan Pike of Alston for the hand-forged ironwork, to Nikki Fionda of Newcastle for all the sewing and to Dominic Shannon for everything involving a sledgehammer. Thank you to Jennin Bernard, Marilyn Warren and Nicola Rose, none of whom made it for the group photograph, but who were an integral part of the decorating team. Special thanks to Sabina Rose, who worked all hours with me to get the house completed, and to the rest of the team, Rachel Morris, Heather Phillips and Lesley Thompson. Thank you to our customers, without whom The Stencil Library would not exist. To those of you who have bullied me into writing this book, I hope you like it. Finally, huge thanks to my partner Chips for not running out of ideas, patience or energy while creating our home.

Publisher's Acknowledgements

With grateful thanks to Annabel Grundy who styled many of the shots and who provided props for photography.

First published in 1998 by
B. T. Batsford Ltd

First published in the United States in 1999 by
The Overlook Press, Peter Mayer Publishers, Inc.
Lewis Hollow Road
Woodstock, New York 12498

ISBN: 0-87951-915-0

Library of Congress Cataloging in Publication Data

Morris, Helen 1960 –
The Stencilled Home: 13 themed room styles / Helen Morris.
p.cm.
ISBN 0-87951-915-0
1 Stencil work—Amateurs' manuals.2. Interior
decoration—Amateurs' manuals. I. Title.
TT270 .M67 1998
746.6—dc21

1 3 5 7 9 8 6 4 2

Styled photography by Colin Poole. Step-by-step and materials photography by Janine Hosegood
Designed by DWN Ltd, London

Printed in China

CONTENTS

INTRODUCTION

This book is an attempt to show some of the vast repertoire of effects that stencilling can impart to an interior scheme. Inside, you will find profiles of thirteen rooms from my own home. These are rooms that are lived in on a daily basis, not photographic sets, and as such have to be practical as well as inspirational.

I am always conscious that most people, myself included, may have tight budgets to work to and the advantage of stencilling above many forms of decorating is that it has the ability to transform the most humble and mundane of materials and surfaces.

A popular misconception among many folk is that stencilled decoration begins and ends with the horizontal border. With luck this book will go some way to dispelling this belief. Stencilled borders are important, and have been around for centuries, but they are only part of the story.

I want to inspire you to decorate. I want you to sit in your rooms, look at your surroundings, close your eyes and imagine colours and schemes. This book provides many stencil patterns for you to cut onto card or stencil film and all of the designs featured in the photographs are available from The Stencil Library. Stencilling is enjoyable and easy, it does not need a large wallet, youthfulness or an art qualification. Courage and imagination are helpful, but they will come with experience. Remember, every professional was once a beginner.

Enjoy it!

The
Rooms

Chinoiserie | 1
bedroom

The term 'chinoiserie' loosely refers to the western interpretation of Chinese art and artefacts which were introduced to Europe in the eighteenth century. Although I was born and brought up in the Far East and love the colours and rich ornamentation of these lands, it is the European evocation of the Orient, shown in this bedroom, that is my passion.

THIS GUEST ROOM LISTS AMONGST ITS INSPIRATIONAL SOURCES THE BRIGHTON PAVILION, A WONDERFUL PALACE ON THE SOUTH COAST OF ENGLAND, RESPLENDENT WITH DECORATIVE EXCESS. IT IS ONE OF MY FAVOURITE HAUNTS AND IS WELL WORTH A VISIT.

WALLS

I had always wanted a yellow room in the house. It was ideal to use as a background here and looks good in contrast with the black furniture. Black is a consistent accent colour in this scheme. It appears on all the accessories, and features heavily on the walls by way of two borders, one depicting fretwork from skirting board to chair rail height, and the other lining the top wall edge with a border adapted from a minaret on a Chinese

wall cabinet. Potential stencil designs can be found in a myriad of different objects.

This room has an unusual shape and the fireplace is set into a curved wall which would normally have been the site of a corner. This is the only fireplace in the house to have found favour with me and consequently it was left unaltered. It has large, deep turquoise tiles edged in black and a decent size mantel and grate.

Six different scenes – one including our two cats – are depicted in the wall panels and these are duplicated along each wall. The subjects are loosely based on the style of Jean-Baptiste Pillement, an eighteenth-century French painter, whose whimsical chinoiserie scenes were widely imitated in Europe. The colour of the wall panels was matched with the fireplace tiles and the panels were edged with a band of copper paint. Emulsion paints and acrylics were applied with small paint rollers and the cats were stencilled with a brush to get the detailing exact. The roller method is quick once you get used to it.

FLOOR

The sanded, wooden floor was painted with black emulsion, over which a basketweave design in brick red was applied with a paint roller. This was my first stencilling adventure with rollered paint and with doubt and trepidation, tempered by an impending deadline, I just got on with it. The ground rules are the same as for brushwork: work your paint well into the applicator, get rid of the excess colour and do not press too hard. This is a big room and the floor basketweave took just two hours to complete. Once dry and whilst I was still flushed with my roller success, the borders of waves and the ornate stars were applied. The drama on this floor is supplied by giant Regency dolphins whose opulent metallic hues are integrated with touches of the turquoise colour from the wall panels.

FABRICS

All the fabrics were picked up inexpensively at my local factory store. The yellow fabric used for the head and footboard also edges the bedspread, which is made from finer silk of an iridescent green and black. Colourful, smaller designs were applied to an assortment of velvet cushions and these were trimmed with red tassels. Red is thought to be lucky and the guests in this room can enjoy many accessories in this colour.

FURNITURE

The Chinese Chippendale style of furniture was
reinvented by our joiner to our design and used on the
huge bed dominating the room. We drew the fretwork
design onto paper so that he could trace it with a router
to produce this delicate woodwork. The head and
footboards were painted with black eggshell and backed
with a deep yellow slub silk. The tall headboard conceals
a wardrobe on either side and these are curtained with
the deep yellow; this is quite a good way of conserving
space in a guest room, as the wardrobes have enough
shelf and hanging space to last two people for an
average visit. Underneath the mattress board are several
cupboards on each side to house all that rubbish that
you do not need, but cannot part with. I made a
miscalculation over the bed height and therefore had to

provide steps to reach it. These have matching fretwork
sides to them and now form an integral part of the
design, lending a further air of importance to the bed.
This is an example of how a mistake can sometimes
work to your advantage.

Originally the furniture pieces were either pine or
MDF blanks. They all started their transformation with
primer and coats of red silk emulsion. Decorative
contrasts were provided by trimming in brick red,
overlaid with black, copper and gold and, on completion,
all surfaces were protected with acrylic varnish.

Finding chinoiserie-style furniture to paint for this
room proved difficult. I had not the time to scour the
junk shops and in truth most were selling the same
country-style pine reproductions which we already
stocked in our own shop. So I used a farmhouse pine
table and chair and set about transforming them. A large
stylised chrysanthemum head radiates from the table
centre and the pine chairs have simple star
embellishments.

I saw the rattan chair while hurrying past a shop in
London to catch my train home. It was closing time and
my arms were already full of luggage. I managed to
leave with the chair and a large ceramic lamp in an even
larger box and fought my way onto the tube train in
rush hour. My tip for getting a seat on the tube at peak
time is to take one with you, sit on it and ignore your
squashed fellow passengers who will be incredulous at
your nerve!

The chair had a decent sized panel on the back and
seat. I applied the same trellis pattern used on the
drawers in my kitchen and broke it up by painting a
black, oval shape with a small version of the Regency
dolphin from the floor decoration in the centre.

ACCESSORIES

The table lamps were bought in a sale; their shape was just right, reminiscent of old tea caddies. Originally they were green with an English farm scene transferred onto them. This was soon changed, using the same painting process as was used for the furniture. Delicate linework was applied with a gold felt pen around the two black bands circling the lamp base. I managed an acceptable rendition of scrollwork and it was easier than expected.

Acrylic varnish was used over it, as oil-based varnish tends to melt the ink.

Similar ornamentation to the trellis pattern used for the rattan chair was applied to a bin, a tissue box and a small writing desk in the window. As a final touch, copies of original eighteenth-century engravings by Pillement were displayed in matching black frames, hanging from pink taffeta ties.

STEP-BY-STEP
STENCILLING WITH ROLLERS

This is a quick and easy way of stencilling, and is particularly handy for large jobs. The correct amount of paint is paramount here – too wet and the paint will seep, too dry and nothing happens. Using spray adhesive is preferable to tape with the roller method.

Materials

- Emulsion paints: raw sienna, black, dark brown, creamy yellow, burnt sienna
- Medium size dense foam paint roller
- 4 small foam paint rollers
- Repositioning adhesive
- 5 roller trays or hardboard sheets
- Lining paper

Method

1 Pour a little of each colour paint into roller trays, run the roller through the paint and then roll off the excess paint onto the lining paper. Do not use newspaper for this, as the print will come off. Apply gently but firmly through the stencil. Do not press.

2 A different roller is needed for each colour – smaller ones for details and larger ones for the background.

3 The colours blend beautifully and as long as you do not need very intricate detailing, rollering is fine for the beginner. I keep a hairdryer handy to speed up the drying time, as I tend to get a lot of damp, excess paint on the stencil when using rollers.

4 Clean your rollers thoroughly with warm, soapy water and make sure they are totally dry before you next use them.

Campaign | 2
room

This guest bedroom was the first to be completed. The styles are based on an imaginary campaign tent from the Napoleonic Wars. Its motifs and borders were all fashioned on an Empire theme and decorated with military precision. The Emperor Napoleon was fond of the bee motif, which swarms over the tented ceiling, and he used the five-point star shown on the walls and bed hangings. Other typical Empire motifs are the acanthus leaf border, the leopard skin and the laurel wreath, which are all in evidence here.

MY INSPIRATION CAME WITH A GIFT OF TWO
IRON EX-ARMY BEDS AND A PHOTOGRAPH OF A
BEDROOM IN THE SCHLOSS CHARLOTTENHOF IN
POTSDAM, GERMANY. THE LEAN LINES OF THIS
ROOM, WHICH WAS OCCUPIED BY STAFF
MEMBERS, WERE FASHIONED AFTER THE OWNERS'
CAMPAIGN TENT. THE POTSDAM ROOM WAS
BLUE, WHITE AND STRIPED, THE TWIN IRON BEDS
AND DECORATIVE TRIMMING WERE SYMMETRICAL
AND SPARSE AND EVERYTHING WAS PORTABLE,
JUST AS IT IS HERE. I ALWAYS KEEP A TOUCH OF
RED IN THIS ROOM, WHETHER IT BE A MILITARY
JACKET OR A VASE SPILLING WITH TULIPS. RED
DETAILS LIFT MANY COLOUR SCHEMES AND WORK
WELL WITH THE BROWN SHADES USED HERE.

WALLS AND CEILING

The walls of the campaign room were papered with a
roll of brown parcel wrap from a packaging wholesaler.
Vertical stripes of chocolate emulsion were applied with
stencilling in shades of gold on top.

For the ceiling we bought loads of purple polycotton
sheeting, turned on the music and boogied our way
through the hundreds of tiny gold bees needed to adorn
the tented ceiling. There was a panicky moment after we
had sent all the fabric to the flameproofers. When it
came back, parts of the gold paint had oxidised with the
flameproofing solution and had turned the most hideous
shade of green! The positive side to the story is that it
calmed down after a few days – unlike me – and the
new green, gold and sometimes coppery shades of the
bees do add a sort of aged authenticity, as if it has been
hanging there for years. As the weather gets damper the
bees go greener and then for a few weeks in the summer
they are their originally intended gold again. The fabric
was lifted into place, and then pleated and stapled onto
wooden battens around the walls. I still have my fabrics
flameproofed whenever possible, because we are visited
by the public, but I now stencil it after treatment.

CARPET

The carpet was decorated with a wide frieze of Greek
keys, following the contours of the room. A large,
centrally placed garland of laurel leaves in brown,
overlaid with a black leopard skin design, forms a focal
point and this was sprayed on with aerosol car paint. I
find that if you apply the stencilling, vacuum over it and
then repeat the process, the aerosol patterns will last a
long time. I used cheap, corded carpet, but the frieze
gives an impression of bespoke carpeting which would
have been above my budget.

FABRICS

Bed canopies were made from mattress ticking and were
lined with cream sateen. I discovered that Napoleon
insisted on fifty stars per square metre. I ensured that
this was the case here, and they shine above the
occupants of each bed, helped by pin spot lamps fixed
onto plaster corbels. The bedspreads were a lucky find.
The original plan was to stencil covers for the iron beds,
but when I found an 'end of roll' at my local fabric shop
which echoed both the rope and the acanthus border
already used in the room – and it just happened to be
perfect for colours – I felt there was no point in making
unnecessary work.

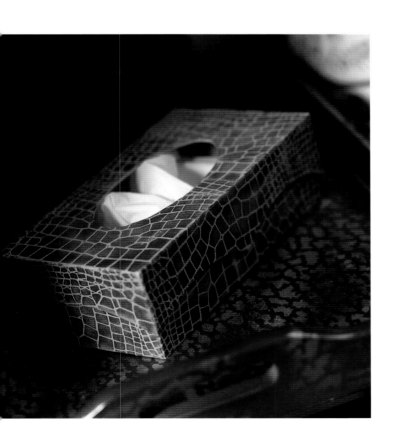

ACCESSORIES

Instead of a wardrobe, two sturdy metal arms were
attached to the wall for hanging clothes. Beside them
stands a large empty wooden trunk and a collection of
storage boxes. These are made with papier mâché and
are stencilled with animal prints. A plywood box was
given a marquetry effect by stencilling empire motifs
with woodstains. The box sits on top of one of a pair of
leopard print butler's tables.

The back of the door was given an opulent
tortoiseshell paint effect and the tatty plastic doorknob
was gilded with gold metal leaf. The same leaf was
applied to the window sill and the gold dulled with a
layer of pale brown wood varnish.

STEP-BY-STEP
HOW TO USE STENCIL-TEX TEXTURE SCREENS

Stencil-Tex was used in the Campaign Room to create the leopard skin stripes and crocodile effects. These intricate brass stencils are made by The Stencil Library. They can transform the simplest stencil shapes into something more sophisticated, or alternatively give pattern or texture to a flat surface.

Materials

- Burnt umber acrylic paint
- Stencil-Tex (crocodile)
- Large stencil brush
- Low tack tape or repositioning adhesive
- Kitchen paper

  Note: to combine crocodile Stencil-Tex with the stencil shown in the photograph, use yellow oxide acrylic and another stencil brush.

Method

1 The technique for stencilling with Stencil-Tex is no different from standard stencilling techniques (see pages 126–137), but ensure that the paint has been well worked into your brush. The brush should also be even drier than usual. Stencil-Tex can be used on its own (as here) or with another stencil.

2 Stencil-Tex is used to embellish the Harlequin Diamond shape (see page 74). First of all, stencil the shape in yellow oxide and then sandwich the Stencil-Tex between it and the background. Then apply the second colour, burnt umber, which is a good crocodile colour. When the Stencil-Tex is revealed, you will find a diamond pattern of brown crocodile with a yellow background.

Blue and white | 3
dining room

Inspired by eighteenth-century blue and white china, this dining room was modelled on Georgian simplicity. A deliberately plain, glass-topped table reveals the patterned floor below. Stencils of plates, pots and teacups grace the walls in crisp formation. This is where my students eat on our courses and all family dinners are held. It is simple and friendly by day and, with the help of dimmer switches and dozens of candle lanterns, it is glamorous by night.

Stencils with many overlays have been used in this room, resulting in visitors mistakenly concluding that the images were cut from paper and stuck to the wall. Using multi-layers or overlays is similar to printing, in that you use a different stencil for each colour that you apply. The images become more elaborate as each layer is applied. A great deal of stencilled pattern in historic buildings is built up in this way. It is common to see this used on the walls and ceilings of churches.

WALLS

The stencilled images of plates, cups, teapots and coffee pots have between two and five layers and are arranged formally within panels defined by a narrow band of mushroom-coloured paint with copper stencilling overlaid. There are twelve different blue and white images and each is separated from its neighbour by a little honeysuckle motif, also used in the kitchen next door. A trellis and leaf effect border sits at the top of the walls on a dark blue background; the blue provides a perfect complement to the metallic copper and aqua paint. The designs were mainly based on eighteenth-century English porcelain patterns.

FLOOR

The sanded wooden floor was painted in the same dark-blue emulsion which is also found in the kitchen. A square stencilled 'carpet' forms a centrepiece, flanked on all sides by the trellis, stencilled again in copper and aqua, bordered each side by a narrow stripe of putty-coloured paint. A dinner plate design bearing pheasants and peonies appears in each corner. A huge circle made up of four quadrants of ribboned flowers was applied around the centre of the floor and frames a larger pheasant plate 125 cm (50 in) in diameter. Everything was stencilled by brush using emulsion and acrylic with generous amounts of a copper colour to give warmth.

ACCESSORIES

The room is not cold or overlooked, so curtains were never considered. A large fireplace with an oversized mantle dominates one wall – its tiles which were once brown have been given a facelift with tile primer, blue paint and then satin varnish. A carved wood surround displays several glass hurricane lanterns and some wonderful fake cabbages bought from my own shop.

The chairs are rustic and were originally chapel chairs; there is a shelf on the back designed to hold your prayer book, but now they usually store each person's napkin. I was going to paint them and finished one as a test piece, but the effect looked too contrived, so the others were left in their natural shabby state. The hand-forged tables were made by the friend who makes all my iron items. Apart from flowers, food and candles, no further additions were needed.

STEP-BY-STEP
MULTI-LAYER STENCILLING

Many stencil companies sell overlaid stencil designs. As a general rule, the first stencil should have at least two registration holes cut into it.

Materials

- Stencil
- White, dark blue and light blue acrylic paints
- 3 medium or large stencil brushes
- Low tack tape or spray repositioning adhesive
- Pencil
- Kitchen paper

Method

1 Place the first stencil in position and stencil through it with the lightest colour. Before removing it, put a mark through the registration holes with pencil or chalk; you can do this directly onto the wall or fix a small patch of low tack tape onto the wall underneath the registration hole and make the mark on that.

2 Then apply the light blue. Each layer of colour is applied using a different stencil, all registration

points will fit over the marks that you made through the first stencil. Continue until all the layers are used.

3 After removing your last stencil overlay, which is usually the darkest, remember to peel away the pieces of tape or erase your pencil marks.

First layer (pale colour)

Second layer (darker colour)

$\mathcal{R}$aj | 4
room

Every house I have owned has had a red and green room; I find this combination warm and comforting. The inspiration for this room originally came from some attractive leather-bound art journals whose well-worn covers were decorated with gold and black toolwork. It only became known as the Raj Room after the acquisition of a pair of limed wooden howdahs. How anybody made long journeys on an elephant's back in these chairs defeats me. Making the comfy velvet covered seat pads became a priority.

THERE ARE FEW STENCILS USED IN THIS ROOM, BUT THEY ARE APPLIED TO EVERY WALL AND MOST ACCESSORIES. THE MOTIF FLOWS THROUGH ALL THE ROOM'S ELEMENTS, BUT THIS IS NOT OVERPOWERING, AS ALL SURFACES (EXCEPT THE PERSPEX LIGHTING PANELS) ARE UNIFORMLY STENCILLED IN AN ANTIQUE GOLD WITH NO SHADING OR HIGHLIGHTING – JUST FLAT COLOUR, APPLIED WITH A BRUSH. GOLD STENCILLING WORKS WELL WITH MOST COLOUR SCHEMES AND MANY STYLES. DEPENDING ON HOW IT IS USED, IT CAN GIVE SIMPLE OR GLAMOROUS RESULTS.

WALLS

The impression of old leather books on the walls is produced with an easy paint technique. The walls were painted with an emerald green vinyl silk and divided by bright red panels in the same finish. A chalked plumb line and spirit level will create the panels (see page 136).

To achieve the leather effect, several glass jars of oil glaze were made up in varying shades of dark green to black to go on the emerald background, and burgundy to brownish black for the red panels. A 60/40 transparent oil glaze (also known as oil scumble glaze) to white spirit was used for the mixture. It was tinted with artists' oil colours. Taking the emerald walls and ceiling first, patches of different green to black glazes were randomly applied with cloths, leaving a few small areas untouched, and then the colours were merged by patting over the surface with a big pad of muslin, or any lint-free, soft fabric. The idea is to blend the edges of the glazes together. The harder you press with your cloth, the more the background paint will show through, giving you brighter areas. After the green glazes had dried, the edges of the new green panels were protected with strips of low tack tape. The same treatment was given to the red panels with burgundy, brown and black glazes. The tape was removed and any fuzzy patches were disguised where the glaze may have leaked under the tape. Gold stencilling was applied up to the dividing line, camouflaging the join between red and green panels.

The same stencils were used in our school room, only this time the panels are a dusky pink and lilac colour washed over a cream background. The stencilling is also cream and the result is sugary and pretty; my students are often surprised to discover that it uses the same scheme as the Raj Room. This shows how changing the colours can lend a whole new lease of life to a set of stencils.

RADIATOR

The radiator was painted in exactly the same way as the walls, and stencilled with a tall floral motif in gold. This was a very fast stencil job as the photographer was on his way when I decided to tackle it and the radiator was so hot that it fried the paint and my fingers. An explanation of radiator stencilling is given in the Arabian Nights chapter (see page 94).

FURNITURE AND FABRICS

The joiner made the bookcases to our design, the bottoms of which are illuminated by lightboxes, sometimes used in bars for displaying wall menus. A couple of fluorescent tubes are set behind a translucent material – in this case a cream-coloured perspex. This was stencilled using car spray paint through the same floral motif as the radiator. The perspex was attached to an MDF frame cut to match the radiator. The light shines out through these panels, highlighting the stencilling and giving a soft background glow.

I could not afford the kind of expensive sofa needed for this room, so I made an alternative. Delightfully squashy, it converts to a bed by removing the cushions. A plywood platform was made and on top sits a mattress with a black velvet cover similar to a fitted sheet. Below the frame is a large storage area and in between is a fringed expanse of stencilled velvet, falling over the arms of the frame, held in place by Velcro. Fat padded cushions, home-made and bought, vie for space with our animals. The gold paint used throughout the fabric

decoration was our own make and, as long as you 'set' the paint with an iron, it is perfectly washable. The silk cushion covers were a sale purchase and a delight to stencil. We bought over a dozen of the pink ones and they appear in a few of our rooms. Velvet takes a little more patience, but it will work with a dry brush method (see page 128) and spray adhesive. This sofa has no back, so here it backs on to the façade of a church organ. The organ front was planned as a bedhead, but the ceiling was too low to accommodate it. However, it offered a perfect focal point for this cosy sitting room. The seat pads for the howdahs were cut to fit from foam blocks and covered with black velvet. Each side was stencilled, so that when the cushion wears it can be flipped over.

The large teak coffee table had an accident on its journey and arrived damaged, so I paid half its price, painted over the cracks and stencilled it. Buying the Georgian fire surround was extravagant, but necessary, because the existing one was small and ugly. The lamps were a sale find, but the stark light from their white glass shades did not suit, so we sprayed the fronts of the shades with aerosol car paint which threw the light behind them, illuminating the wall pattern.

STEP-BY-STEP
STENCILLING ON FROSTED GLASS

Acrylic frosting varnish is usually a device to obscure viewing, but in this case it was used as a decorative detail to create a frosted effect on some glass door panels. Follow the manufacturer's guidelines for cleaning your decorated glass.

Materials

- Stencil
- Acrylic frosting varnish
- Foam sponge
- Scrap paper
- Saucer
- Low tack tape

Method

1   Firstly, make sure that your glass is thoroughly clean. I wiped mine with vinegar and water. Then position your stencil using a low tack tape. You can use a stencil brush and dry paint method, a roller, or a foam sponge (the one shown here). Firstly the corners of the sponge were gathered in the hand in order to make a firm applicator. The frosting varnish was then decanted into a shallow saucer and the sponge was dabbed into it. The sponge was blotted onto scrap paper to remove the excess varnish, then gently tapped through the stencil.

2   Acrylic frosting varnish takes a little longer to dry than acrylic paint, so wait about 30 seconds before repositioning.

Note: If you wish to disguise an undesirable view, then you can cut small shapes from stencil film, stick them on temporarily and then frost the surrounding area. The film is then removed leaving only small clear shapes to view through.

Honeysuckle | 5
kitchen

This kitchen was originally two rooms and a corridor. As we had to rewire and replumb, we decided to take the walls down too and have a kitchen to party in. Our landlord had suggested removing the blue, solid fuel stove; I declined and designed the layout and colours to complement it. It repays me by constantly demanding feeding and belching coal dust into the room, and consequently there is little in the way of accessories in this room, quite simply because anything which would attract dust is kept behind doors. For this reason, visual interest has to be provided by diverse decoration of the many surfaces.

After the renovation we were left with a large, disjointed area with two differing floor levels and six doors leading off from it in all directions, so there was still much work to be done to make it both a pleasant and practical room. The kitchen is dark for most of the year – its windows are small, deep and few, so we set halogen lights into the ceiling to compensate.

WALLS

The lower level walls were painted in Regency white and the upper level a dark blue. Something was still needed to unify the space, and we decided upon a honeysuckle design which, by its rambling nature, would seem to embrace walls and kitchen surfaces and gather the areas together. The honeysuckle grows from each of five large oriental pots, stencilled in dark blue and ochre, which appear to balance on the skirting board of any area of wall large enough to accommodate them. Stems weave around each other, spreading their branches over the

ceiling, onto the window sills and into the recesses of the many doors. Soft, blended stencilling with acrylic paints was used; there were four colours in the honeysuckle and two in the pots. I was keen to achieve a vaguely oriental theme for the kitchen and the honeysuckle fits well with this idea. There are also oriental scenes on the cupboard doors surrounding the stove and a stencilled panel of a Mandarin couple advertising their wares on the opposite wall.

FITTINGS

Further continuity in the kitchen scheme is provided by a delicate tracery design, stencilled in aqua and copper over door panels, drawer fronts and parts of the fridge. This kind of stencil is very useful for any size of area. The pattern is just repeated until the space is full. Some of the panel decoration has been broken up by a centrally placed stylised flower. This flower appears sprigged over turquoise silk, which takes the place of cupboard doors

under some of the work surfaces. The same flower is present in the dining room – as both rooms are next to each other and the door is usually open, it ties the two rooms together. Another design also used in the dining room is the deep trellis border at the bottom of the silk curtaining. The teacups stencilled in the dining room are again represented in the glass paintings that appear throughout the kitchen. The idea of using fabric in place of some cupboard doors is another budget one. Attached by Velcro, the fabric opens and shuts just like any other door. A cheaper alternative to silk could be cotton.

The fridge is the most unusual feature of the kitchen. It sits on top of a large painted desk with the legs chopped off. The desk is big enough to provide a

seating area around the fridge, which in turn is housed inside a stencilled wooden case with shelves on either side. The fridge doors have been disguised with sheets of plywood, fixed in place with strong carpet tape, and stencilled with dragons adapted from wallpaper seen in the Brighton Pavilion. An alternative idea would be to decorate the white metal doors with car spray paint.

All the work surfaces in the kitchen are made from MDF, painted with vinyl silk and protected with varnish – firstly with acrylic varnish and then finished with two coats of oil based varnish. The tiles were plain, white and cheap. They were decorated using a paint, specially intended for ceramics, glass and most shiny surfaces. Once the pieces are decorated, they are baked in a normal oven for around 40 minutes at a medium heat. Whilst cooking that batch of tiles, I made plates and bowls to match. These paints are available from many stencil or art shops.

STEP-BY-STEP
USING FREE-FORM STENCILS

The beauty of free-form stencils is that you grow and mould them to suit your purpose; they are perfect to twine around beams, sloping ceilings and recesses. A similar design is given on pages 48–49.

Materials

- Stencil
- Acrylic paints: yellow oxide, Baltic green, dark Victorian rose, raw sienna
- 4 stencil brushes (medium and large)
- Spray repositioning adhesive
- Kitchen paper
- Lining paper

Method

1   Start the honeysuckle by putting in a general framework of twigs. This is not a strict rule and you can start with the flowers if you prefer, fitting in branches and twigs afterwards. When stencilling the leaves and flowers, apply the palest colour first and the darkest last. This stencilling used a total of four colours.

2   Then attach leaves and flowers to your twigs – we used the brush, acrylic paint and colour blend method outlined in Stencilling with Paint on page 128. Try to avoid the whole thing looking too twiggy and be extravagant with the flowers. It is sensible to practise first on the lining paper until you get a feel for the different components of this design.

Nautical | 6
bathroom

THIS BATHROOM IS PURE THEATRICAL FANTASY. AS THIS
WAS THE LAST ROOM TO BE COMPLETED AND BECAUSE WE
HAD BEEN SO CAUTIOUS WITH FINANCE THROUGHOUT THE
DECORATION OF THIS HOUSE, WE BLEW THE BUDGET ON THE
BATHROOM AND CREATED AN EPIC! FIRST WE WOVE A
STORY AROUND AN ANCIENT MERCHANT VESSEL SAILING
ALL THE WAY TO THE FAR EAST IN SEARCH OF EXOTIC
PLANTS AND RETURNING WITH AN EXQUISITE HAUL OF BLUE
AND WHITE CHINA, AND THEN WE SET ABOUT CREATING
THIS IN OUR DECORATION.

It took two stencillers three days to complete this job; the joiners had the largest task and that is where the budget went. Apart from the elements detailed below, finishing touches were prints, plants and the haul of blue and white treasure. Then all that was needed was to fill the china with pretty soaps, fill the bath with perfumed water, fill my head with pretty thoughts, light the candles and drift away.

WALLS AND CEILING

Navigating charts would be needed for the journey, so the existing textured wallpaper was painted with emulsion and the 6.75 m (22 ft) wall was lavishly transformed into an ancient sea chart, resplendent with galleons, mythical beasts and sea gods. To navigate by the stars, the heavens are mapped out on the ceiling, and prints of palms in flat wood frames remind the sailors of their quest.

Most of the wall decoration revolves around four small compasses, topped with a fleur-de-lys motif. These link like satellites, connected by lines of felt pen around a huge and ornate parent compass. With a diameter of 2 m (7 ft) and every inch of gold hand gilded with Dutch metal, the compass took most of a day and a lot of patience to complete. I swore I would never attempt one again and then agreed to reproduce it for a client's floor the following week. Surrounding the four compasses are several sea gods and mythical beasts. These, and the calligraphy, were stencilled in black. Making the calligraphy stencil took some time, but is a sensible option if your freehand skills are limited – it is much easier to erase mistakes when drawn on stencil film rather than the wall.

A black border skirts both the wall and ceiling, defined by stripes of red and beige. Bright gold painted lettering, enriched by the addition of gilding powder, announces itself in Latin to visitors. On the remaining five walls of this extravagant room, the bubbly textured paper and prolific pipework were covered by plywood sheeting and a 2.5 cm (1 in) wide moulding to give the appearance of panelling.

FURNITURE AND ACCESSORIES

Shutters were commissioned, furniture purchased and floorboards sanded. Everything wooden was dyed with acrylic wood stain, until all shades of the different woods were fairly uniform. On top of that went a further wood stain, a home-made recipe of dilute blue and grey emulsion paint. When dry, it was limed by applying white gloss paint thinned with turpentine. The mixture was wiped onto the wood, allowed to settle for a minute and then most of it was removed with a dry cloth. This concoction smells evil and devours your rubber gloves! The liming was used to imitate the effect of years of salt water on wood, evoking the feeling that all is sea-washed. Copious amounts of dead flat varnish were used to dull all the wooden surfaces.

The linen chest was a prototype for the first decorated furniture that I made. Stencilled with a multi-layered design over a blue and black colour wash, it proved so expensive that I only sold two. The problem of blending a shower curtain into the surroundings was overcome by making two hinged concertinas of wood, forming shutters on each side of the bath when extended. They fold flat against the wall when not in use.

A coffee table and Portuguese steamer chair are not on everyone's list of vital bathtime requirements but as the space was there, we have indulged ourselves. Scented candles, soaps and fresh white flowers, displayed in pots of blue and white porcelain and silver bowls, bedeck the table's limed surface. A large cupboard sits by the sink, housing all the necessary tubs, bottles and towels.

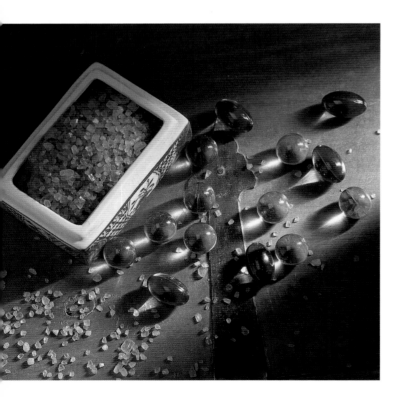

Four giant wooden buttresses were bolted into place and appear to hold up a painted night sky, emblazoned with hundreds of pale blue and white stars on the ceiling. The stars orbit a large disc of astrology signs stencilled with copper, gold and aqua, mimicking the verdigris on the many candle lanterns. The star stencil was cut into small pieces so that it adhered easily to the ceiling. We attempted to do the same with the astrology disk, but got the measuring wrong and had to do some serious 'filling in' to camouflage our error.

STEP-BY-STEP
STENCILLING WITH DUTCH METAL LEAF OR GILDING POWDER

Acrylic gold size and a stencil brush are needed for either of these alternatives. Acrylic gold size is a glue that is milky when applied but turns clear in seconds. To see where it has been applied, add a little acrylic paint to it. Size is runny; make sure you work it well into your brush by scrubbing it on kitchen paper and drying the bristle tips (see page 128).

Materials

- Acrylic gold size
- Acrylic paint (to colour the size)
- 2 large stencil brushes
- Soft brush (such as cosmetic brush)
- Gilding powder
- Dutch metal sheets
- Kitchen paper
- Repositioning adhesive

Method (for gilding powder)

1 Stipple or swirl the mixture gently through your pattern and remove the stencil before the size sticks. Let the size become touch dry with a light tackiness – usually between five and twenty minutes.

2 With a soft, dry brush, work the gilding powder over the stencilled area. It will only adhere to the size, so a piece of paper will catch the excess powder which can be dusted off with a dry cloth and then reused.

Method (for Dutch metal leaf)

1 Place the transfer metal leaf metallic side down on the tacky gold size, press it firmly into place and peel off the backing. The metal leaf will stick to the size.

2 The leaf may not peel away neatly, so finish off by using a soft, firm stencil brush over the gilded area. Use the excess leaf to patch up areas. Alongside the metal leafed stencil is the finished design in gilding powder.

Panelled | 7
room

People who saw the panelled room in its original state are always amazed by the way that we transformed it. We had envisaged a room with an Elizabethan/Jacobean feel – not a strictly authentic one, but achieving something between the museum and the movie set.

THE ORIGINAL GOOD POINTS ABOUT THE
EXISTING ROOM WERE TWO LOVELY WINDOWS
WITH GARDEN VIEWS. ITS BAD POINTS WERE A
TINY, BROWN TILED FIREPLACE WHICH THREW
OUT NO HEAT AND THE WOODEN SHUTTERS WERE
WELDED OPEN WITH LAYERS OF YELLOWING
GLOSS PAINT.

WALLS

First of all the walls were clad with plywood sheeting
and battens made from 2 x 1 timber. Our joiner added
simple detailing to the panel edges with a router and
then we painted both the panelling and the battens with
black emulsion. Off-white emulsion was applied to the
ceiling and around the top of the room and a bold black

frieze was stencilled. This kind of strapwork stencil is
very quick to execute with a large brush, and a minimum
of skill is needed as the pattern is very easy to match up.
Any smudges made when squashing the stencil into the
corners of the wall were quickly rectified using a slim
paintbrush. The radiator was painted to make it
disappear into the wall. Into each of the wood panels
went either a small stencilled crest or one of twenty-
two different flowers. The inspiration for the flowers
came from a folio of Tudor flowers at the Ashmolean
Museum. The flower panels were stencilled brightly with
the name of each flower appearing underneath it. They
were then dirtied by the application of an oil-based
varnish with a large dollop of raw umber oil stainer
mixed in to suggest years of ageing. To finish, little
rosettes made of polyurethane were glued between each
panel and covered in the same black paint.

FIREPLACE

The fireplace was a stroke of luck. We had looked at several surrounds, but a friend suggested that we buy nothing until he had taken a sledgehammer to the area. He demolished the first fireplace, followed by a second, and then found a third, which came complete with roasting paraphernalia and this one we decided to keep! Another friend made the iron grate, fire dogs and the many candelabras to our design. One of the lovely things about having a huge log grate is that you do not have to clean the grate but can allow the ash to pile up throughout the year, otherwise it would burn too fiercely. In fact the fireplace is so enormous that the fire needs to be lit hours before it is required, to let the heat permeate and the smoke clear. Unaware of this problem, the first time we used it was one Christmas Eve; the last of the rosettes were stuck in place, cushions zipped and fire lit just an hour before our house guests arrived. On their arrival the room was a total smog. Brave friends and family sat low on the floor whilst the smoke blanket swirled above their heads and escaped through the open windows. It was a surreal evening, but we were determined to christen the new room. The aroma of candle wax and wood smoke pervades this room from one winter to the next.

FLOOR

The floor was stencilled with acrylic and then darkened by wiping diluted black emulsion over it. A Tudor rose design sits in the corners, each connected by a line of diamond shapes, flanked on each side by an unbroken line. Sisal matting was used for the floor covering. While this is a wonderful surface to stencil, one of the things about stencilling is knowing when to stop and, in this case this was before we got to the sisal matting.

ACCESSORIES

The dresser was left to us by a relative and sits perfectly here under a collection of pewter plates. Each section of wall carries a simple iron candle sconce on it. The candles cost us a fortune but the sofas were inexpensive enough to afford a pair. We let the dripping wax from the multitude of church candles pile onto the varnished floor. When the wax pile impedes movement around the sofa, we take a swift pastry scraper to it.

Skirts were made to dress chairs. Tables were clothed and cushions covered, all in cream canvas. Then all of these were stencilled. The cushion covers were decorated by using some of the floral designs from the wall panels. When used on this light background, unaffected by stain and varnish, much more of the detail is seen. All the stencilling was applied with brush and acrylic paint. The coffee table was made of chain mail, an essential element for a 'Jacobethan' room.

STEP-BY-STEP
APPLYING HOMEMADE WOOD STAIN

As a precaution, I usually add a coat of matt acrylic varnish to newly sanded boards. This means that if any spillages should occur, they can be easily wiped. When wood is left unprotected, spillage tends to sink straight in.

Materials

- Stencil
- Black acrylic paint
- Black emulsion paint
- Acrylic varnish
- Extra large stencil brush
- Rubber gloves
- Cloth
- Paint kettle
- Water
- Kitchen paper
- Repositioning adhesive

Method

1 With a large stencil brush, paint roller or spray paint, apply your decoration (see pages 128–129).

2 To get the effect of dark wood onto the yellow pine floor of the panelled room, you can wipe a mixture of 50/50 black emulsion paint and water onto the floor. Using a cloth or old T-shirt, soak in the mixture and then wring it out. The cloth must only be damp not wet, so wring it out thoroughly and work quickly.

3 When dry, apply several further coats of acrylic or oil-based varnish and your floor is finished. An oil varnish will add a slight yellowing to the floor. To get an impression of how the boards will look with acrylic varnish, just wet the wood and the colour it turns will be close to the finished result.

Foxgloves

Lady Cap

Parisian | 8
c h i c

THE STENCILLED PATTERN OF DIAMONDS AROUND THE ROOM FORMS THE FOUNDATION OF THE OVERALL DESIGN. DECORATION OF OTHER SURFACES IS KEPT TO A MINIMUM AND THE CONTRASTS ARE PROVIDED WITH TEXTURE RATHER THAN WITH COLOUR AND PATTERN. DOWNY VELVETS, SLIPPERY SATIN, POLISHED WOOD AND HAND–FORGED IRON TEAM UP WITH TERRACOTTA POTS, CREAMY CANDLES, SILVER LEAF AND CRYSTAL TO MAKE THIS SENSUAL BEDROOM THE MOST DELIGHTFUL RETREAT.

WHEN WE FIRST SAW THIS ATTIC ROOM, IT HAD BEEN NEGLECTED FOR SEVERAL YEARS. BUT THE VERY PARTICULAR CHARM OF THIS ROOM WAS IMMEDIATELY EVIDENT, WITH ITS DRAMATIC GABLES, TINY WINDOWS AND VIEW BEYOND.

WALLS

The first stage was to coat the crumbling plaster walls with white matt emulsion. They were then bathed in a colour wash of a soft, taupe-coloured glaze through a harlequin pattern stencil. The colour was wiped through the stencil using rags. Taking the pattern only halfway up the wall lent an illusion of height to this very low ceiling and solved the problem of negotiating the stencil around the multitude of beams and eaves. The moss-coloured velvet which was used to cover the chair perfectly complemented the walls. It was a gift from the famous American actress who, among her other skills, has an inventive talent for interiors.

FABRICS

The first stage of stencilling this simple yet stylish monogram was to secure the fabric with a piece of stiff card which was cut to the shape of the cushion cover. The card was then lightly sprayed both sides with repositioning adhesive and then inserted between the layers of fabric. Once the fabric is smoothed over the tacky card, it will stay perfectly still and will also prevent any paint spillage from running on the fabric layer below. The initials were stencilled with a burnt umber paint colour, using the dry brush and paint method (see page 128).

ACCESSORIES

The original colour of the lamp base was altered to fit in with the bedroom style by using an aluminium transfer leaf which is inexpensive and easy to use. The leaf is available in copper, silver and gold as well as other metallic shades. You will find that the process of leafing is very addictive. The lamp base was first coated with acrylic gold size, a kind of glue for gilding, and then left to dry for about twenty minutes. Once slightly tacky, the aluminium sheets were laid. They measure about 15 cm (6 in) square and need to be handled with care as they are very thin and fragile. They adhere fiercely to the size, but any tears can be repaired by filling in with spare bits of leaf. Once completed, the base was coated with acrylic varnish in case it tarnished. This method was used for the wastepaper bin and several storage boxes.

The paper lampshade was stencilled with silver paint using a squiggles design. When stencilling objects with curved surfaces, try not to be tempted to hold all of the stencil at once. You only need contact with the surface that you are stencilling at that moment. Once you have finished one section, do not remove the stencil but press down the next bit and release the stencil from the area just completed. Continue rolling the stencil in small sections until the design is complete. Note that this method is not effective when spray painting.

DRESSING ROOM

Just off the bedroom, through a low arch, is the dressing room. Although similar in colour to the bedroom, the style here is bold and humorous. Much of the storage space is provided by second-hand luggage, painted to match horizontal stripes on the walls. Plain wooden boxes from our own shop, stencilled in crocodile skin, sit next to the real thing found in antique shops. The crocodile stencil is one of our most popular designs. It was used in the Campaign Bedroom (see page 20) and instructions for its use can be found on page 24.

ACCESSORIES

Stencilled lettering appears on many of the accessories and you can find something witty or pithy to read on every wall. I think the quote above the door to the bathroom sums up this little gem of a space, 'Elegance is good taste plus a dash of daring'.

Stencilling letters can be a time-consuming business. If there is little to be written, I usually design the entire sentence onto stencil film and cut the whole thing in one piece. If it is going to be a large project, I put

straight horizontal chalk lines on the wall (see page 136) and match them up with lines drawn onto the stencil under each letter. This will keep everything straight. I always use a translucent stencil film for lettering, so that I can see through it to line up the letters with each other. To stop the stencil getting clouded with paint, I keep the relevant solvent handy. Keep the space between each letter uniform as well as the large space between each word. These letters were 10 cm (4 in) tall. I allowed 1.3 cm ($^5/_8$ in) between the letters and 5 cm (2 in) between the words, but you can judge the distance by eye depending on the size of the letters, remembering to stick to the same distances. Rather than reaching for the ruler every time you need to position a letter, I found it handy to put a permanent pen mark 1.3 cm ($^5/_8$ in) before each letter and match the mark on the stencil with the end of the previous letter stencilled. Plan and measure out your scheme carefully. It is frustrating to run out of space before you finish your sentence.

STEP-BY-STEP
COLOURWASHING THROUGH A STENCIL

Colourwashing through a stencil with large windows such as the harlequin diamond is easy and effective. I have used this method with many combinations of colour and am always pleased with the results. For subtlety, choose colours similar in tone. For a striking partnership, I like the result of colourwashing brick red through the stencil onto a background of egg yolk yellow. To achieve the look in the Parisian bedroom, follow this recipe. This scheme uses an enlarged version of the diamond repeating pattern (see page 74).

Materials

- Stencil
- 50/50 acrylic scumble glaze and water
- A generous dollop of taupe acrylic paint and a squirt of Payne's gray
- Repositioning adhesive or low tack tape
- Soft lint-free rags
- Rubber gloves
- Bowl for mixture

Method

1 First of all combine the scumble glaze and water mix with the taupe acrylic paint and the Payne's gray. For a room this size, make about three-quarters of a pint (250 ml).

2 Fix your stencil to the wall with the repositioning adhesive, making sure it is straight. Soak the cloth in glaze and wring it out so that it is damp rather than wet. Wipe onto the stencil surround and then into the stencil windows. Always work into the shape and not against it.

3 When you are ready to move the stencil, gently peel it away and line up your stencilled diamond shapes with the drawn lines on your stencil. Check that everything appears to be lined up correctly and then start wiping the glaze on the next section.

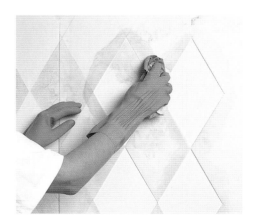

A B C D E
F G H I J
K L M N O
P Q R S T
U V W X
Y Z &

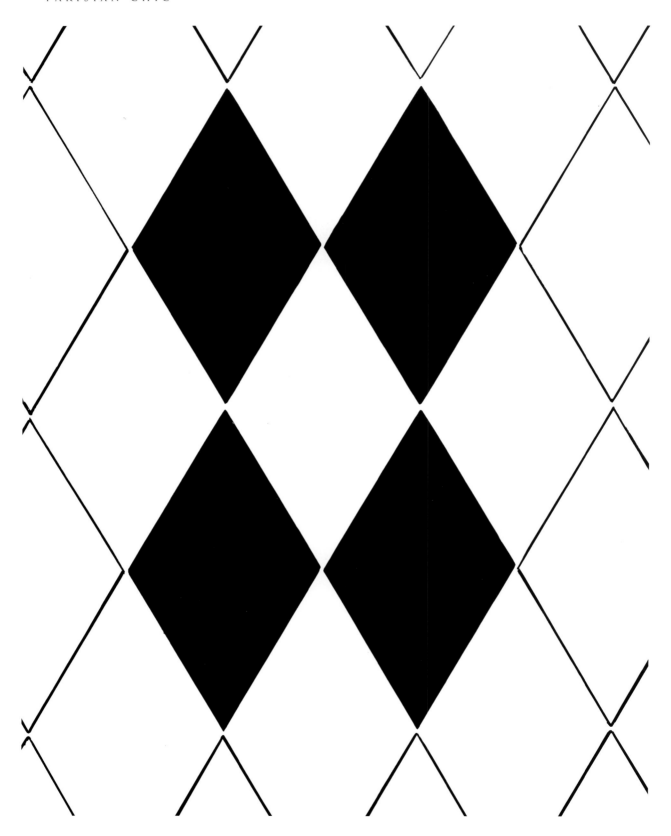

Grecian | 9
hallway

The stairs and landings in this house are always a joy to me. They are south facing and therefore are usually warm and sunny. One of my favourite spots to sit in the house is the window seat on the first floor landing. We have used this scheme, and variations of it, many times and in many situations – it seems to suit all ages and styles of home.

THE DECORATION OF THE HALLWAY WAS KEPT
COMPARATIVELY LOW-KEY SO THAT IT DID NOT
DETRACT FROM THE DRAMA OF THE ROOMS
ADJOINING IT. THE BLACK, WHITE, GOLD AND
GREY SCHEME TAKES YOU ALL THE WAY
THROUGH THE DIFFERENT FLOORS OF THE HOUSE.

WALLS

An anthemion design was chosen for the walls. This is a
stylised flowerhead, probably a honeysuckle, and it was a
very popular decorative image in both Ancient Greece
and Greek revival. The smaller motifs were set up in a
grid formation, eliminating the need to measure again
when repeating your design. I used a swirling motion

(see page 128–9) and a large brush with gold paint, and
I could not believe how little time it took to stencil this
design. Greater time was allotted to painting on the
dado rail which is a black strip with gold lines created
each side with masking tape and gold paint. To add a
little more gold, dots were added at strategic points. The
large anthemia took much longer, because each motif
had to hang straight. There was much muttering and a
heavy reliance on chalked plumb lines (see page 136). It
took nearly two days to get the large motifs stencilled
(there are five flights of stairs), but only one to complete
the small repeating designs.

CARPET

The carpet is cheap needlecord and was stencilled using
household emulsion paint. If this was done again, I would
use my usual concentrated acrylics which are softer and
more hard-wearing. However, with a tin of black and a
tin of white emulsion, the carpet was stencilled in the
same fashion as any of the other projects, although it
needed two layers of paint. On each stair tread is a white

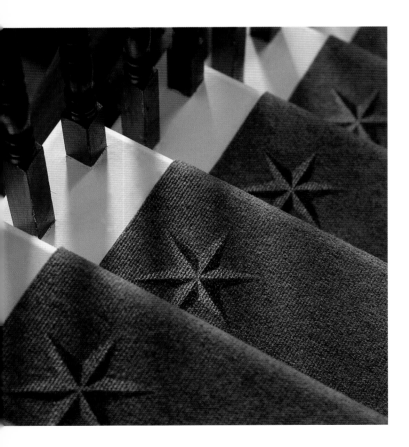

star with a 'drop shadow'; the star stencil has a separate overlay for its shadow. Each landing has an elaborate, stencilled rug, using classical Empire designs that follow the contours of that particular area. This rug is wearing reasonably well, considering that this is the main thoroughfare to everywhere in the house.

LIGHTS

The lights were made locally by a woodturner and their painted lamp bases were then stencilled - this was the most fiddly job you can imagine!

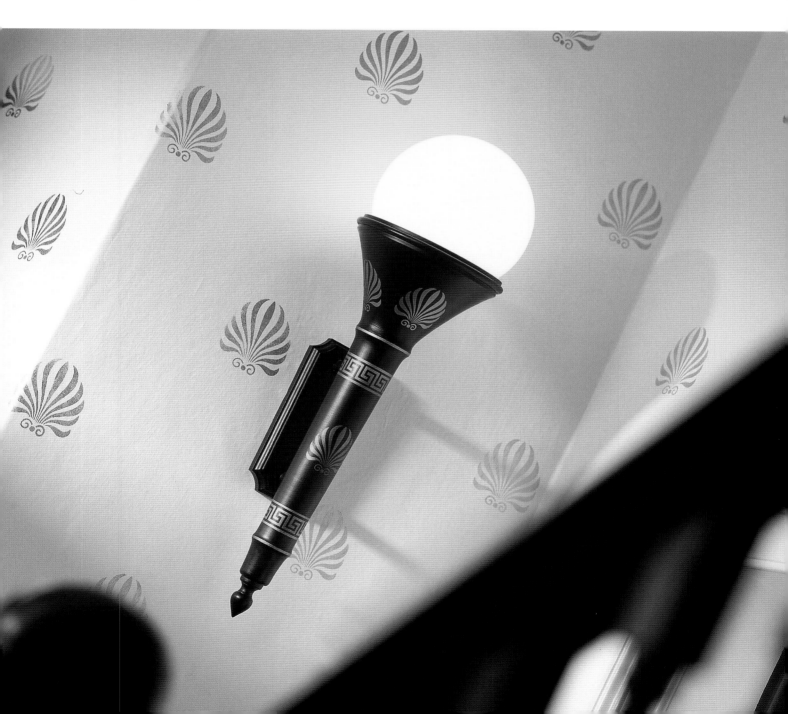

STEP-BY-STEP
STENCILLING REPEAT PATTERNS

These repeat patterns are one of my favourite
stencilling techniques, as large areas of decoration
can be completed quickly. To make a stencil similar to
the one used here, you should trace the anthemion
motif on page 81 onto transparent stencil film four
times. The spacing between the motifs can be as
dense as you require, but the spacing must be equal.
Cut the designs with a scapel or heat pen.

Materials

- Stencil
- Gold paint
- Black acrylic paint
- 1 large and 1 small stencil brush
- Plumb line
- Spirit level
- Repositioning adhesive
- Kitchen paper

Method

1 If working with a square or oblong shape, refer to the
instructions on page 136 to find the centre of the
wall space. By placing the centre of your motif over
the exact centre of your wall space, you will ensure
that your stencilling is symmetrical. You should check
this with a plumb line and spirit level. All your
pattern repeats stem from this one, so if you get it
right, the rest will be easy.

2 Firstly stencil all four motifs. Use your large brush for
the gold and the smaller one for the black. Once this
is done, you only need to use two of the motifs to
continue the pattern; the other two fit over
previously stencilled shapes to ensure that the
spacing is regular.

Note: the stencil shown on page 98 will be a useful
guide for spacing repeating designs.

Gustavian | 10
room

Several years ago I discovered a wonderful set of photographs depicting a selection of eighteenth-century Swedish interiors. I stored the images in my head and when I first stepped into this room, I knew that I had found the perfect canvas on which to paint them. The virtue of two long windows facing south and east provided the necessary light needed for my interpretation of the Gustavian style. This term refers to the style popularised by King Gustav III of Sweden during the latter half of the eighteenth century.

THE PROPORTIONS OF THE ROOM ARE PLEASING, AND IT HAS THE ADDED QUIRKINESS OF HAVING TWO DOORS LEADING TO SEPARATE AREAS OF THE HOUSE. I IMAGINED A HAZY PICTURE OF FADED GRANDEUR; A BLEACHED WOOD, CREAM-CANDLED HAVEN, WITH BREEZY MUSLIN SWATHED AT OPEN WINDOWS. THIS WAS REFLECTED IN A HUNDRED CRYSTAL DROPLETS HANGING FROM ANCIENT CHANDELIERS AND MIRRORED IN SILVERED GLASS WITH CRACKED, GILDED FRAMES. THE MOST IMPORTANT ELEMENT, HOWEVER, THE STENCIL DESIGNS, KEPT ELUDING ME.

In reality my bleached, blonde floorboards screamed peroxide yellow instead of platinum and had to be quietened by a wash of dilute palest pink emulsion. I only managed fifty-three crystals, and the chandelier was from a chain store. Breezy muslin did not keep the sun from blinding me at 4.30 on a June morning, so heavy fabric blinds hang in their place. My Gustavian vision had now become a little fuzzy around the edges.

WALLS

A flash of inspiration reminded me that I already owned the perfect set of stencils for this room. The flowers used on the wall panels had been in The Stencil Library collection for many years, designed after seeing the floral murals at Wallington Hall in Northumberland. Four different flowers appear on the wall – hollyhocks, irises, poppies and foxgloves; the hollyhocks were also used on the closet doors and the irises in the bedspread. The bedspread was given an Empire-style panel to match the walls and was stencilled with the same paint.

The lovely thing about having windows facing both east and south is that the sun rises through one, and the midday warmth soaks in through the other. To colour the woodwork, I chose the exact green that matched the surrounding fields on a June evening and hoped that the colour would still suit in November. In the event, it could not have worked better; it is early November as I write – the green is still perfect and the burgundy and russet colours on the wall panels and blinds are echoed in the neighbouring trees and hills. The rich cream of the walls and the putty colour of the ceiling and panels bear little resemblance to their colour chart swatches, but by a happy accident they look even better than I had planned.

I enjoyed being artistic with the flower panels. There was plenty of scope to blend and shade with the paint giving a soft, delicate finish. Concentrated acrylics and several large stencil brushes were used. Into each flower went yellow ochre, red oxide and olive green, but I varied the rest of the colours so that no two flowers are alike, yet all complement each other. I have used red oxide on the edges of the leaves and yellow towards the centre. There are a large number of leaves and this technique gives more interest than flat colour would.

FLOOR

A floral repeating pattern was stencilled over the floor, the linen chests and hat box. When stencilled in a similar colour to its background, the design gives a believable imitation of a damask fabric and when stencilled with earthy colours it resembles both American and European folk art, so it is extremely versatile. I varied the colours of the pattern on each different surface to avoid the room looking too co-ordinated. Each surface was stencilled with acrylic paint.

The floor was sanded, stencilled, sealed with matt acrylic varnish, then wiped with a diluted pale emulsion paint before finishing with dead flat oil varnish to give a pale, powdery look. My set of prints came from an old calendar and these were a starting point when choosing colours for the flower panels. As a finishing touch, my crumbling stone urns were brought in from the garden, providing a contrast in colour and texture.

FURNITURE AND ACCESSORIES

The bed was supplied blank for painting and placed to get the best of both views over the valley. The wardrobe was built from MDF. The hat box is made of sturdy card and painted with emulsion. The silk plants were borrowed from my shop and have stayed.

I had deep sills made for the windows, which double up as desks and usually have lamps and floating candles on them. Bronze and glass carriage lanterns are grouped over the hearth wall. The mirrored plaster candle sconces were too golden, so they were wiped over with pale, plaster-pink emulsion to dull them.

Once I had finished the main elements of this room, I labelled the contents of the cupboards with an interpretation of a Swedish hand-painted alphabet. As soon as the word 'shoes' appeared in that lettering, looking suitably Scandinavian, I knew that the room was now close enough to my daydream to allow me to collect my clothing, cats, hats and husband and move in.

STEP-BY-STEP
STENCILLING A WALLPAPER REPEAT

This pattern was used a great deal in the Gustavian Room with a method that I call 'random swirling'. It involves two or more colours (Baltic green and bronze yellow in this case) which are swirled together. It is a fast and effective way of incorporating accent colours into your decoration scheme and saves time differentiating between leaves and flowers or whatever your chosen design happens to be.

Materials

- Stencil
- Yellow and green paints
- 2 large or medium stencil brushes
- Low tack tape or repositioning adhesive
- Kitchen paper

Method

1 This stencil joins together to make a wallpaper effect, so make sure that your first one is positioned correctly, as the pattern will grow from it.

2 After finishing your first pattern, fit the printed registration areas over the completed work. They must fit exactly, because a minor error at this stage could become a major one ten feet into the pattern. If this does happen then do not panic, but just edge the pattern back on track a little more with each application until it appears to be correct again, and then proceed more carefully. Crayons or paints are perfect for executing wallpaper repeats such as this one. Spray paint is less suitable, because the registration patterns, vital to repeating your stencil correctly, get obliterated. If you wish to use spray, mask over your registration lines at each application.

Arabian | 11
nights

This room feels very much like a Hollywood film set and is fun, opulent and outrageous. Less restrained than some of our other rooms, this one cannot claim an unpainted surface. A major advantage here is that, because of the two walk-in wardrobes, we did not have to sacrifice any valuable wall space for storage.

THE MORNING SUN FLOODS INTO THIS LAVISHLY DECORATED BEDROOM. THE ABUNDANCE OF GOLD AND COPPER PAINT ON EVERY SURFACE HERE MAKES THE RICH INTERIOR OF THE ROOM GLISTEN — ON A SUNNY DAY THE EFFECT IS DAZZLING.

WALLS

The main wall in this room is really quite splendid; unbroken by doors, window or furniture, it allowed us to decorate the surface with a series of large panel designs. The designs for this room were drawn up with mathematical precision and we made a scale plan to follow. There was a lot of measuring involved, but the

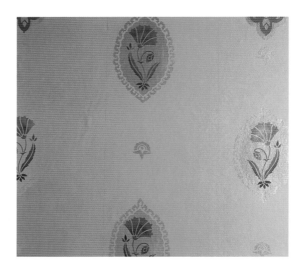

actual stencilling was mainly a three-colour job and was comparatively quick to do.

Four of the panels depict the same vase spilling over with flowers and between these appear slimmer panels showing a tall tree. This is stencilled on a russet-coloured background with green, vermilion and gold. All these panels are separately bordered by a trim line of flowerheads, and the whole display is set into a frame of frilled petals and flowing leaves on a black background. This wall took four stencils and two days to complete.

The other three walls evoke the feel of a fabric tent. The dark brown colour of the ceiling ends on a fringed border, overlaid with burgundy and copper. Below that hangs the second layer on an area of terracotta which is embellished with lozenges of gold with poppies on them. Each row of poppies faces in a different direction and the pattern ends on another deep, fringed border below the chair rail. On the area representing the third layer of fabric, regimented rows of copper motifs shine from their pale pink background. All the woodwork is painted in black satin finish, except for the floors.

RADIATORS

The black, painted radiators had large poppies of
vermilion, gold and green stencilled over them. Radiators
are not a job for the faint-hearted. The first part of the
stencil was fixed in place with repositioning adhesive
and each part was pressed to the contours of the
radiator as work progressed, freeing the completed bits
of stencil, so that it gradually rolled its way in and out
of the grooves. The radiator was then enclosed in a
cabinet of MDF with carved windows to enhance the
decoration and allow the heat to escape.

FLOOR

When stencilling floors, I always tie cushions to my
knees - it looks ridiculous, but makes the work more
comfortable by protecting me from the hard floor. We
started off with a really poor quality wooden floor which
splintered badly. We gave it several coats of the wall
paint and that seemed to bind the wood until we could
drown it in varnish. The background of this floor was
painted burgundy and was edged in the same border as
the panelled wall. Once the main background colour was
dry, the tile pattern of black hexagons was laid out and
stencilled. Inside each of the tile shapes, either an

abstract floral design was stencilled in gold with red and green, or a pattern of golden grapes with purple brushed over it. Several coats of matt acrylic varnish were added over the next few days.

CEILING

The difficult thing about stencilling a ceiling with a large design is that gravity decrees that the stencil tends to spend more time in your hair than in its proper place. To minimise this, I used two thick cardboard 'arms' to hold the stencil rigid. These arms were about 45 cm (18 in) long and were attached to each other at right angles. The stencil was vaguely triangular in shape and measured over three feet long. The arms were taped to the two long sides of the stencil which, in turn, was

taped to the ceiling. Unfortunately removing and repositioning the stencil caused the card to buckle and before long the wet stencil had draped itself on my head. If repeating this job, I would construct heavy-duty 'arms' from wood, attach them to the ceiling with sticky Velcro pads and repair the damaged paintwork afterwards.

FABRICS

Stencilling fabric is usually a delight and the bed dressings were no exception. When stencilling lengths of fabric, I have the luxury of several yards of table space. The tables are sprayed lightly with repositioning adhesive before smoothing the fabric on top. This holds the fabric securely, allowing accurate stencilling.

ACCESSORIES

The sturdy card hatbox used as a side table was coated with emulsion paint to match the walls and then stencilled with designs which had been previously used in the room. It was finished off with acrylic satin varnish. The bedside lamp has been leafed with Dutch metal, a cheap substitute for gold leaf. The lamp was originally rusty metal which was cleaned, primed and coated with gold size. When the gold size dried to a tacky finish, sheets of the thin gold were laid on, with the excess leaf rubbed away where the holes were. The halogen bulb set inside throws a wonderfully patterned light over the various surfaces of the room.

The doors and the fireplace used the same repeating pattern as the velvet bolsters on the bed. The fireplace tiles were stencilled with acrylic paint and then protected with a coat of varnish. A plain, papier-mâché tissue box holder and wastepaper bin were stencilled over the same background colours as the room, using the lozenge design and the poppy. Touches like these are inexpensive but complete the overall style of the room.

When the fabric is lifted up, the next piece is moved into position, smoothed down and the process starts again. If you don't have a large surface area, a kitchen work surface or a wallpaper pasting table will do just as well.

The bedspread was stencilled using copper, antique gold paint and a very bright gold mixed from gilding powder and fabric medium. This makes a beautiful paint, but it does go off quickly once made. The bedcovers took an entire day and evening. Stencilling the green velvet for the bolsters was easy. After sticking the velvet to the table each pattern repeat took just seconds to do, applying copper paint and a squiggle of red.

The bed canopy was made from a turquoise iridescent silk. Stencilled with a two-layer stencil, the first layer was copper and gold and the second was burgundy, viridian, brown and scarlet. The hangings were suspended from a gold leafed broom handle, hung from ornate cup hooks in the ceiling with a dressing-gown cord. The other side was attached with Velcro to a wooden batten on the wall.

STEP-BY-STEP
STENCILLING ON DARK SURFACES

This method involves undercoating the stencil with a lighter colour, and can be applied to many surfaces. The following demonstration shows the stencilling of the silk on the bed canopies and uses burgundy, veridian and gold water-based paints. You can undercoat your stencilling with white, which is fine, however my preference is for gold, as I love its warmth and softness. You can obliterate the gold with your overlaid colours, or you can allow it to show through to provide highlights. When using fabric paints, follow the manufacturer's guidelines.

Materials

- Stencil
- Gold paint (suitable for walls and fabric)
- Viridian and burgundy acrylic paint
- 3 stencil brushes (1 large, 2 medium)
- Repositioning spray
- Kitchen paper
- Tailor's chalk (to colour registration marks)

Method

1 Lightly spray the work surface with repositioning adhesive to hold the fabric in place. Stencil the gold undercoat first – apply a total coverage through your stencil, swirling the paint on with a large dry brush. As usual, use a different brush for each colour.

2 Apply the next colours wherever you want them using the remaining two brushes. Remember that without the gold underlayer, the dark colours will become lost when applied to a background of similar tone. Mark the registration points with tailor's chalk, rather than pencil.

3 Remove the stencil to reveal the design.

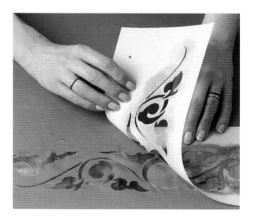

Robert Adam | 12
study

At the bottom of our garden, there grows a wild flower orchard, which is a joy for most of the year. This is where you find our 'garden shed', or Robert Adam Study, nestling under the branches of an aged pear tree and positioned so that it gets the sun from dawn to dusk. The humble exterior makes it all the more delightful when you open the door and step into a glittering miniature roomscape, illuminated by candles whose light dances on the gold surfaces and creamy white stencilling. Designed to imitate the intricate plaster panelling in many of the masterpieces engineered by the eighteenth-century architect, Robert Adam, this interior relies on formal symmetry for its impact.

THE INTERIOR OF THE SHED WAS CLAD IN
HARDBOARD, TO GIVE THE FLAT SURFACE
ESSENTIAL FOR THE FORMAL PANEL WORK, AND
AS A BONUS THIS PROVIDED INSULATION. I LOVE
WRITING BY CANDLELIGHT IN THIS ROOM, AND
SITTING HERE SNUGLY, CONTENTED AND WARMED
BY COFFEE, I WELCOME THE FIRST WINTER SNOW
IN THE GARDEN.

WALLS

First of all the walls were painted with an emulsion of a
biscuit colour. Over that, the wall space was broken up
with circular and rectangular panels of sky blue paint
and a stripe of dark blue was applied about 30 cm (1 ft)
deep along the top of the walls. The stencilling, which is
just flat colour, was carried out using dense foam rollers
and white acrylic paint. Onto the dark stripe went a
stencilled cornice of scroll patterns and anthemia. The
rectangular panels all feature flowers, flowing from
fanciful urns, and the circular panels display musical
instruments. The gentleman featured above the desk was
intended to be Apollo, God of the Sun and Patron of
Music and Poetry. In fact he turned out to be Hermes,
the Messenger of the Gods, so this room loses a little of
its symbolism – you can guess who did not pay attention
during classics lessons!

CEILING

The ceiling of the shed slopes gently towards a pitch
roof, its two halves separated by a batten, embellished
with a pattern of Greek keys running the length of it. A
large classical motif, accompanied by a fine swag and
rosette, was placed symmetrically on each slope. The
anthemion motif, similar to the one used in the Grecian
Hallway, was one of the principle classical motifs
appearing in Ancient Greek and Roman artefacts and
architecture, and symbolised a flower. During the
eighteenth century, the anthemion became enormously
popular and was used extensively by many artists,
architects and decorators, including Robert Adam.

FLOOR

The floor was painted dark blue and a stark, white Greek key design is stencilled around the edges. A varnished canvas carpet known as a floor cloth was made to fit just inside the stencilled border. These floor cloths were popular in Europe and America until the invention of linoleum and other easy-wipe floorings. Floor cloths are versatile, hard-wearing, portable and, best of all, can be created to complement their surroundings totally. This floor cloth makes an attractive and wipeable protection against muddy gardening boots and was easy and inexpensive to make. A length of medium-weight canvas was bought from a flag maker. Buy a size to suit your requirements plus a 5 cm (2 in) hem allowance on each side. Both sides of the canvas were primed with leftover odds and ends of emulsion paint. Three coats of midnight-blue emulsion then were rollered over the front of the cloth, followed by one coat of matt acrylic varnish. The matt varnish presents no problem to stencil over and makes spillages simple to clean up.

The expanse of blue was broken up with a circle of claret-colour paint. To draw a large circle like this, use a length of wood batten with drilled holes, large enough to take a pencil, at 5 cm (2 in) intervals. A slim nail was fixed at the other end and a pencil was placed in the hole which would give the desired radius. The nail point was held at the centre of the circle and the pencil end of the batten was then rotated to draw the outline of the circle. Once completed, the circle was painted freehand, following the pencil line. A large gold 'H' was centred on the cloth, and all the decoration grows from that point. The 'H' was wreathed in a circlet of laurel leaves and a slim gold border of laurel defines the cloth's edges. The chalked string method (see page 136) was used to get a straight line for the hem, and to mark the lines for the laurel border on all sides of the cloth. The hem was positioned 5 cm (2 in) from each edge, and the line for the border 15 cm (6 in) inside that. The straight edge of the laurel stencil was simply butted alongside the inner line.

All the stencilling in this room is simple to do, except perhaps the mitred corners on the floor (see step-by-step sequence on page 105). When the decoration of the floor canvas was complete, the hem allowance of 5 cm (2 in) was turned on each side. This can be aided by scoring along each fold line with blunt scissors, ensuring that the canvas is not cut. Then the corners of the hem allowance were trimmed across to achieve a smooth result when the hem is folded. The hem was glued with PVA and then weighed down until bonded. Several coats of acrylic varnish were applied to the front of the finished cloth for protection. Although the cloth will feel flimsy at the outset, it will become quite weighty after the application of paint and varnish layers.

ACCESSORIES

A canvas director's chair, a gilded MDF desk, a writing tray and candle sconces are the only accessories in this room. All have been given an undercoat of red oxide primer and are leafed with gold Dutch metal. The primer could have been any contrasting colour, but red oxide is most commonly used. The distressed look on the gold was achieved by dipping a little wire wool into methylated spirits and rubbing gently over the leaf to expose the undercoat. The gilding was not protected here, because it was felt that if the gold tarnished during the winter, then it would still suit its surroundings perfectly.

STEP-BY-STEP

HOW TO MITRE A STENCILLED CORNER

You will need to use the chalked string method (see page 136) to mark the border lines on each side of the cloth, and butt the straight edge of your stencil alongside it. The ideal mitre tool is simply a paper rectangle or square. Fold the square diagonally in half, or fold the rectangle to make a perfect 90-degree angle.

Materials

- Gold paint
- Stencil brush
- Tape measure
- Length of thin string
- Chalk
- Paper
- Low tack tape or repositioning adhesive
- Kitchen paper

Method

1 Stencil the border and, as you approach a corner, place the paper with the diagonal edge crossing the point on the floorcloth where the two chalk lines meet at the corner. The straight edge of the paper should be exactly parallel to the chalk line on the floorcloth's edge that you are not currently stencilling. Stencil your border up to the diagonal edge of paper, and then remove both the paper and the stencil.

2 Reposition the paper to fit over the diagonal area you have already stencilled, and reposition your stencil with the edge on the next chalk line. Then begin stencilling the border in its new direction. When you remove the stencil and paper, a neat corner should appear, similar in appearance to the mitred corner of a picture frame.

Contemporary | 13
bathroom

The shape of this rooftop bathroom suggested a tented theme for the decoration and the colour combination of pale turquoise and pillar-box red was inspired, quite literally, from a dream. The dominant red colour together with hot pink and gold accessories bring a much needed touch of heat. Such warm colours are essential to this room, as three of the walls are external and the draught through the skylights can be enough to blow dry your hair!

THE COLLEAGUE WHO DREAMT OF THIS COLOUR SCHEME DESCRIBED WEARING A RED DRESS IN A TURQUOISE SPORTS CAR AND LOVED THE VISUAL IMPACT OF THESE UNUSUAL COLOURS. NEITHER OF US HAD EVER SEEN A REAL EXAMPLE OF THIS COLOUR COMBINATION, SO WE SPENT TIME COMPARING BLUES AND REDS ON PAINT CHARTS AND FABRIC SWATCHES. THE BLUE WAS EASILY RECOGNISED BUT THE RED WAS ELUDING HER. EVENTUALLY WE CHOSE TWO REDS FROM A SHADE CARD AND HAD OUR PAINT STORE MIX THEM TOGETHER.

WALLS

The walls were measured and the stripes were calculated so that all walls were symmetrical and no stripe went into a corner. This was worked out with a calculator, by measuring the walls and then dividing the space by the desired width of stripe. You need to have an odd number of stripes on each wall, so adjust the width of your stripe as necessary. A width of 30 cm (12 in) was the closest we could get and it was still necessary to stretch a few stripes to 33 cm (13 in) to make it work. With a fat stripe such as this, it is possible to take a few liberties without them being noticed. There are seven stripes on the short walls and thirteen on the long ones.

FLOOR

A centrally placed trompe-l'oeil bath mat was painted in blue and edged with a gothic border in gold. The rest of the floor was covered with an eight-point star design and all was finished with several coats of eggshell oil-based varnish. The bath was sprayed with gold spray paint and has red stars stencilled over it.

CEILING

Visitors often ask how the strong red colour wash was achieved on the ceiling. This was simply a result of losing the recipe for our specially mixed red emulsion. Instead, we supplemented what little paint we had left with acrylic glaze which we then brushed on and hoped for the best. Two morals can be drawn from this - have the courage to experiment or always remember your paint recipes!

FABRIC TENT

Gold touches most surfaces in this cheerful room but is most spectacular on the fabric tent. Tents like this make an interesting alternative to cupboards or wardrobes. These can be bought ready-made, or can send your own fabric to the suppliers who will make the tent and post it back to you complete with the frame in kit form (see page 143). We had a beautiful iridescent silk in dark aquamarine and wanted a vibrant, gold fabric to complement it. None could be found so we made our own from a pale rose gold gilding powder mixed with fabric medium and totally saturated the material with it. Its incandescence is a perfect foil for the silk. The tent hides all those useful bathroom things that have no artistic merit. The silk exterior curtains a series of shelves and wire baskets. A large swirling Paisley design begins at the hem and gives way to little flowers, stencilled in gold to emulate sari fabric. Flowers are stencilled in jewel-coloured acrylics on the gold painted trim of the tent.

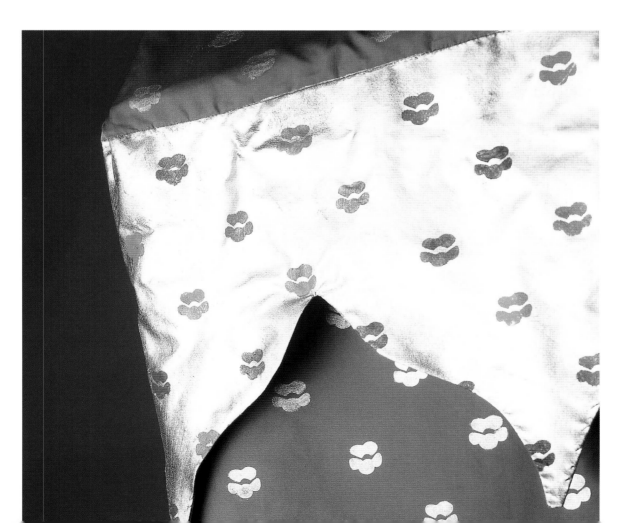

ACCESSORIES

The bathroom is large enough to afford the luxury of a chaise-longue, resplendent with cushions of a hot-pink colour which were picked up in a sale and customised by stencilling them with smaller versions of the designs used elsewhere in the room. Hot-pink and red is one of my favourite colour clashes.

A great deal of effort was spent trying to find inexpensive but interesting bathroom lights. We didn't manage to find anything suitable at the right price, so we settled on cheering up a cheap and acceptable frosted glass lantern by stencilling it with a gold emblem borrowed from one of the cushions.

The storage units were plywood filing cabinets. The drawers were removed, reversed and a hole drilled through what was now the front. Tassels were applied as drawer pulls, ensuring that a strong knot was tied behind each one. Some clear cut-glass door handles in an assortment of colours were attached to anything remaining that would require opening. When painting wooden boxes or furniture, I often use a foam pad or stencil sponge to apply the base coat. Quite simply dip it in the paint and wipe. The colour will go on thinly and smoothly, without brush marks. To get a transluscent look, rather than opaque coverage, the sponge should be dampened with water first. The thin stain sinks straight into the wood and is touch dry within minutes. If the colour is too weak, wipe another layer over it (see photographs on top of facing page).

STEP-BY-STEP
MAKING YOUR OWN GOLD PAINT

Gilding powder mixed with metallic medium makes a rich paint with a brilliant lustre. Depending on the medium used, it is often washable on fabric. The down side of this paint is that it must be used fresh, as within a few days it can turn solid in the pot.

Materials
- Metallic medium or fabric medium
- Gilding powder
- Empty jar
- Stirring stick

2   This bright stencilling is a combination of a pale rose gold powder mixed with fabric medium.

Method

1   Put the medium into a jar and sprinkle enough gilding powder to form a thick, gooey paint when stirred. Make sure the powder is thoroughly mixed in.

Materials | 14
and techniques

You can start stencilling with the minimum of equipment and after having mastered just a few basic techniques. The joy of stencilling is that it is so simple yet so effective. You can transform a room or just a cupboard quickly and easily, creating subtle or dramatic effects.

MATERIALS

The number of specialist shops catering for stencillers and paint effect enthusiasts is increasing. Some department stores will hold a few items in their craft or home decorating areas. Mail-order companies offer a good choice and are convenient to use.

There are various applicators and paints associated with stencilling. Some of my customers apply their paint with velvet pads, others with sponges and rollers and one just uses her fingers. My personal preference is for a concentrated acrylic paint applied with a dry brush method (see page 128). This paint is washable on fabric and sets fast onto walls, floors and furniture. It even makes a plausible wood stain when diluted.

My basic starter kit would consist of a small, medium and large stencil brush, plus a few colours. I find the most useful ones are yellow ochre, deep red and dark green, followed by a deep blue and gold. This small selection is suitable for many subjects and period styles.

Stencils

Stencils are tools for producing accurate repeating patterns and they are usually made from a water-resistant material. Waxed card and metal stencils are available, along with ones made of thick plastic. The most popular choice appears to be a translucent polyester film. The film, available from many stencil specialists, has the advantages of being washable and flexible and is always my choice when drawing, cutting or applying stencils.

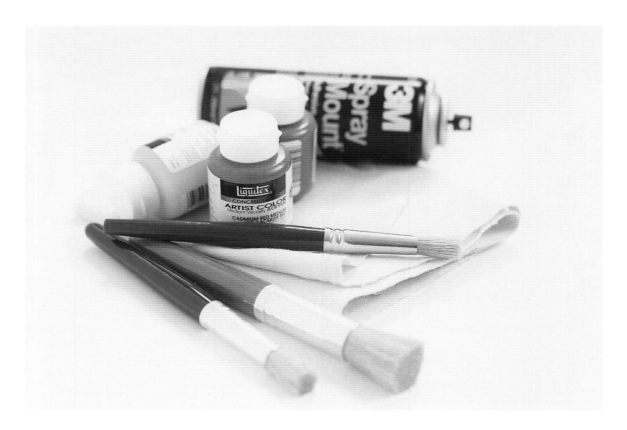

Oiled or waxed stencil card is becoming more difficult to find. I think an intricate design cut well onto card is a beautiful sight. A polyester stencil, for all its technical superiority, does not stir the same feelings. However, I stopped using card stencils years ago, as I can find no advantage in their use.

Metal or brass stencils are hardy and can last for years. An unsurpassed degree of intricacy can be achieved in these stencils, because the designs are etched out with acids and no actual cutting is involved. Devotees of paper embossing favour metal stencils to create greetings cards and stationery. I find a greater degree of skill is needed to stencil through them. If using paint, your brush needs to be even drier than usual. Metal stencils are neither flexible nor transparent and are probably not the best choice for a first timer.

Your stencil should have registration points either printed or cut into the stencil. This will take the guesswork out of stencilling repeats (see page 72), allowing your pattern to flow without visible joins. Motifs or solo designs will not have these marks, as they are intended to stand alone.

Background Paints

I normally use matt or silk emulsions for a base colour; matt provides a preferable surface for working on. Whichever you decide on, make sure that you keep some base paint spare in case of mistakes.

Stencilling Paint

You can stencil with most water-based paints, but you will get a better result more easily if you use stencil paint or a liquid concentrated acrylic. Many are washable on fabric.

Oil Paint Sticks or Stencil Crayons

Stencil crayons are oil-based paints in stick form and are suitable for use on walls, wood, fabric, paper and plaster (see pages 130–131). They are not suitable for use on gloss paints or shiny surfaces.

Spray Paint

Spray paint – either car spray paints or acrylic enamels – can be used for stencilling. The colour range is not vast but they blend beautifully.

Brushes

For the best results you should use proper stencil brushes. These are not expensive and do help to achieve professional results. My ideal brush would be flattish, round and have flagged bristles (split ends) and the bristles should be long and tightly packed, but flexible. Use bigger brushes for larger areas and small ones for details. When dropped on its face, a good stencilling brush should have a slight bounce. This makes it easier on your arm when undertaking a marathon stippling session. The bristles of the brush should also be soft enough to stencil silk without pilling the surface. Ensure that your brushes

are cleaned with an appropriate solvent after use. The Stencil Library manufactures a brush scrubber, which is a square of plastic covered with sharp teeth. This combs all the dried paint from your brush within seconds.

Foam Rollers or Pads

These are an alternative to using brushes and will allow you to cover large areas of flat colour very quickly. Use them with roller trays and emulsion paint (see page 17).

Self-heal Cutting Mat or Glass Surface

A self-heal cutting mat is the best surface on which to cut out a stencil and these are available from office supply or stencil shops. Alternatively, cheap greenhouse glass is a more economical option. If using a heatpen to cut your stencils, use glass as a cutting surface. I would recommend a 35 cm (14 in) square piece of 6 mm (¼ in) horticultural glass. Use one side for the heatpen and the other for blade cutting. Tape the raw edges of the glass with thick tape, as they are very sharp.

Scalpel or Craft Knife

Used for cutting out stencils. Make sure that it has a sharp blade, as dull ones are difficult to control and can be dangerous.

Heatpen

This electric tool melts out your stencil shapes and is very easy to use – just trace the outline of the design with the nib. You must cut on glass when using a heatpen, as it will melt a cutting board. I have also heard that heatproof cutting mats will soon be available.

Kitchen Paper

Kitchen paper is perfect for creating a wad on which to circle a loaded brush of paint and drying the tips before applying it. Avoid using tissue as it shreds.

String and Weight

This piece of equipment is useful when marking straight lines.

Wallpaper Lining

A roll of wallpaper lining is handy for making stencil proofs, so that you can experiment with your chosen colours and techniques.

Spray Repositioning Adhesive

I prefer to use spray repositioning adhesive for securing a stencil. It is also the best option if your stencil is very delicate or floppy. The spray is also my first choice when using aerosol paints or when stencilling fabric. If the stencil is very large, you may need to use a combination of spray and tape.

Low tack masking tape

This is an inexpensive, but inferior, alternative to the repositioning adhesive.

Chalk

A coloured chalk is useful for marking surfaces to help position your stencils correctly.

Varnish

You can choose either oil- or water-based varnishes. Flat varnish gives little or no sheen, eggshell or silk is the medium stage, with gloss as the shiniest and most hardwearing. There is rarely a need to varnish a wall after stencilling, but stencilled furniture should be protected. Floors should have at least three coats of varnish, and three more if you have the patience, in order to give them maximum protection.

TECHNIQUES

A selection of the more popular techniques of stencilling are outlined here. It really does not matter in most cases which one you choose. The important thing is that the paint does what it is supposed to and that the end result gives you pleasure.

Planning

Some people can hold a decorating scheme in their head; most need to plan their work. Experiment by stencilling lining paper with your chosen colours and techniques, so you perfect the image before you start. Cut around and save all these rough images to use as measuring tools. If you will be stencilling motifs, taping these proofs in situ will show whether the patterns fit well into the space allowance. These proofs are moved around easily, and allow you to experiment and approve a design before you start work. Stencilled borders should have a similar 'dress rehearsal'.

When planning a room, you may find it easiest to start with a focal point and work from there. Establish your horizontal patterns first, verticals second and any motifs and embellishments after that. Use coloured chalk and string to mark your walls prior to stencilling (see page 136). Vertical lines that travel the entire length of a wall from ceiling to skirting often look more tailored if they are met by a horizontal border at both ends. Always work from the top to bottom of the line.

Preparing Your Surface

It makes sense to give yourself the best possible surface to stencil on – there is little point in creating your masterpiece on a background that is not worthy of it. The surface should be clean and dry and have any necessary repair work completed before you start.

Walls

Many of my walls have been papered with wallpaper lining and then painted with matt or silk emulsion. However, lining paper is not important to have as a stencilling base and a painted plaster wall offers a perfect surface. Decorating oil-based, eggshell walls is also possible, but not as easy. Wood chip and textured paper can be stencilled using crayons or paint, but the stencil will not lie flat on the surface, making spray painting less successful. Printed wall coverings such as colour washed, striped and marbled paper can be stencilled with stunning results.

A matt-finish wallpaper or paint is far preferable for stencilling than a glossy one. However, some glossy surfaces can be stencilled with acrylic paint and a stippling method. If the surface is compatible, try spray paint or acrylic ceramic paint, both of which make good alternatives on a shiny surface.

Floorboards

Newly sanded floorboards benefit from a coat of matt acrylic varnish before decoration. Any stencilling errors can then be simply wiped off, rather than sanded away.

Furniture

When painting furniture prior to stencilling, I use matt emulsion. This gives a good surface to decorate, is quick to dry and also odourless. Previous paintwork should be cleaned, sanded or removed as necessary. Painted furniture should be protected with varnish.

Fabric

Fabric should be laundered before stencilling. Many textiles are sprayed with size, which removes with washing, taking your handiwork with it.

Other surfaces

Tiles and glass should be cleaned and then rinsed with water and a touch of vinegar prior to stencilling. Some artificial surfaces such as melamine will accept emulsion paint after a sanding with fine grain glass paper or a coat of sanding liquid.

Cutting a Stencil

You are free to copy the black-and-white templates in this book for non-commercial use, but most stencil patterns designed in this century are copyright protected. Do not reproduce or photocopy from books, catalogues or magazines where no permission is given.

The easiest way to trace the design is with a piece of polyester stencil film and a permanent felt-tip marker or a chinagraph pencil. Place the film over the page and trace the design. Some stencil films appear milky, but they are transparent and you will see every detail of a design through them.

If you need to enlarge or reduce a design, use a photocopier to copy onto paper and then trace as before. To reproduce your stencil onto card, sandwich a piece of carbon paper between the template and your card, following the design with a pencil using firm pressure.

If your initial attempts at cutting are not perfect, do not worry – I have seen some surprisingly good images stencilled through some poor-quality first attempts.

Method

1 Use a self-heal cutting mat or sheet of glass. Hold a scalpel or a craft knife like a pencil and make your cut smoothly and continuously (see photograph above right). Do not lift your blade more than necessary, always cut towards yourself and manoeuvre the stencil rather than the knife.

2 Cut out all the areas to be stencilled, as well as the registration marks. As you cut the stencil it will become floppier, so start with small details, and work from the middle outwards. If you break one of the stencil bridges, repair it by placing sticky tape over the wound and cut the shape again through the tape.

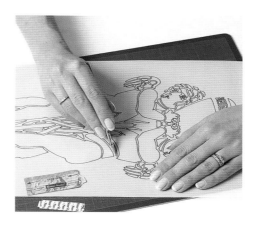

3 A heatpen can be used to cut polyester stencils by tracing the outline of the design with the nib of the pen (see photograph below). The shape should pop out. Do not forget to cut your registration crosses. Cut onto glass when using a heatpen.

Adhering a Stencil to a Surface

In my experience, the best adhesive to secure your stencil is an adhesive repositioning spray. A light spraying will last a long time, without leaving traces of glue on your surface and this is invaluable when stencilling fabric or using spray paint. Always follow the manufacturer's safety instructions. A cheaper, but inferior, alternative is to use low tack masking tape.

Dry-brush Stencilling

Emulsion, acrylic and water-based stencil paint will dry instantly if you use this method. Use a separate brush for each colour and ensure that each one is dry before starting work.

Method

1 Shake up your paint so that it coats the inside of the lid, making a handy palette. Put just the tips of the bristles in the paint and circle the loaded brush onto a wad of kitchen paper. A vigorous swirling disperses the paint evenly across your brush and works the paint up to the bristles (see photograph below). Do one last swirl on a clean patch of kitchen paper to dry the tips of the brush.

2 Before your first attempt at stencilling, do a test stipple on a piece of paper, and run your finger instantly over the mark – if it smudges, go back to your kitchen paper. It is surprising how little paint you need and how far it goes. Stippling, or pouncing, is produced by tapping the flat end of the brush straight up and down onto your surface, keeping your brush at 90 degrees so that the bristles do not escape under the stencil bridges (see photograph below).

3 Another method of stencilling is to swirl or circle the paint through the stencil. The method is just as the names describe – just 'tickle' the paint through the stencil windows. Begin the stencilling motion on the film just outside your design, and swirl your way into the shape (see photograph at the top of the next column). The photograph below this one shows the set of fruit on the right hand side stippled, and the fruit on the left swirled. It adds tone and texture to a one-colour scheme if you use both methods.

Using Spray Paint

You can achieve wonderful effects stencilling with spray paint and they blend beautifully. I use a mixture of car spray paints and acrylic enamels. They can cover most surfaces, including shiny ones such as tin and glass. Some colours will withstand laundering on fabric. Wear a protective mask and ventilate the room to avoid breathing the spray from these cans. Use acetone or nail polish remover to remove car spray.

Method

1 First apply a light amount of adhesive spray over the entire back of your stencil. Smooth it into place ensuring that there are no bits sticking up which will cause clouding under stencil bridges.

2 Position your spray between 7–30 cm (3–12 in) away from the stencil. Start with the lightest colours first and apply a gentle spray for no more than three seconds. If the paint runs you have sprayed too hard. The colours should be sprayed in the general area of where you want to see them and a funnel of folded card should be made to mask off areas of differing colours and to apply details.

3 The photograph below shows the card positioned along the edge of a leaf. This is to add definition to the leaf edge and to lessen the red colour drifting in from the fruit. By spraying the paint onto the bottom third of the funnel, a fine mist of overspray is channelled down the fold of the card, onto the leaf. This method takes a bit of practice, but is much more controlled than aiming at the area and hoping for a direct hit. You will become adept at manouvering the card around your design.

Stencilling with Oil Paint Sticks

This is a very easy way of stencilling for beginners, where the colours blend well. Brushes should be used with these crayons to disperse the colour. Make sure you take a clean, dry brush for every colour. Some crayons are colourfast when washed on fabric. Always do a test piece and wait 72 hours before laundering. Two of the disadvantages with crayons are that the stencilling will not dry totally for several hours and small lumps of oil paint can flake from the crayon.

Method

1 Use a dust sheet to protect surfaces and have kitchen paper handy to keep oily fingerprints at bay.

2 Using a non-porous palette, such as a tile, a polythene bag or the surround of a plastic stencil, scribble colour from the crayon.

3 Then take up the paint into your brush and apply it by swirling it into the paint, then through the holes in the stencil, using either a circling or a pouncing stroke. Build up the colours in layers and do not overload the brush.

4 In order to create shading, add deeper colour around the edges of the shapes.

5 When using paint sticks on a surface that may need varnish, wait three to five days before doing so. Clean your stencil and brush with white spirit.

Stencil Creams

Stencil creams are similar in make-up to a paint stick; in this case just peel away the protective coating and take the colour direct from the pot. Work the paint into your brush, using a non-porous surface as a palette, and apply. Follow the manufacturer's advice for brush cleaning.

Stencilling on China, Glass and Tiles

This is a great way of customising tiles and personalising table and glassware. Many china paints are now dishwasher safe. You can use soft stencil brushes, foam rollers or pads to achieve a good finish. The following project uses a foam pad.

Method

1   Starting with the lightest colour, tip the paint onto the foam and dab off the excess onto scrap paper (see photograph below). With a gentle dabbing, apply the paint to the desired areas.

2   Use a different pad for each colour. Work in a sequence from the lightest colour to the darkest (see photograph at the top of the next column).

3   Then remove the stencil (see photograph below) and leave it overnight to cure. The finished items should be set aside for 24 hours and then either baked in the oven for 35-45 minutes, or have varnish applied (depending on the manufacturer's recommendations), in order to give them a hardy finish.

Shading and Blending

When shading, start with the lightest colour and finish with the darkest. Follow this sequence and the order in which the colours were applied will be easily remembered. You then only need to take extra care in placing the last colour, as mistakes with paler colours can be covered up.

Method

1 Start by stencilling onto the surround and work your way into the holes from the edges in (see photograph below).

2 Round shapes, such as fruit or balloons, benefit from gradually fading the colour towards the middle of the shape. This gives dimension to the shape, acting as a bloom on fruit, a shine on a balloon, or sunlight on a leaf. Yellow has been applied to some of the leaves on this stencil, which will merge with the green and red and make the image appear softer and more colourful (see photograph below).

3 The next colour can now be applied wherever it is desired. The yellow leaves here are lightly covered with green, but the centre of the fruit has been left.

4 Finally apply your darkest shade, which will emphasise any outlines that you want to define. Keep the darkest or strongest colours to the edges of each shape so that the design edges are clearly visible (see photograph below). If you stipple or wiggle your brush onto the surround so that just a few bristles get to colour the shape, you will achieve delicate outlining. Note that here the largest amount of red paint is on the outside, rather than the inside of the stencil pattern.

5 When repeating your shaded design, do not bother to replicate the first one – a similar effect will be better than an exact one.

Measuring a Straight Line

I have found this method to be the fastest way to mark the guidelines on your surface prior to stencilling. You can buy a special tool for marking lines which consists of a length of string embedded in a well of chalk. I often find that the chalk is hard to remove from matt painted surfaces, so I would recommend using a ball of thin string and a box of coloured chalks. To mark vertical lines on a wall, simply tie a weight (such as a door key) to a chalked string. Once a weight is tied to the string, the string will hang straight, ensuring a vertical line.

Stencilling in a Straight Line on a Wall

This method is much easier if you are working with a partner.

Method

1 First of all mark two points of reference at the desired height on either side of the wall. Then, drag a piece of string over the coloured chalk until you have coloured enough string for the entire length of wall

2 Stretch the string over both reference points and pull it tightly (see photograph at the top of the next column). If you have no partner, firmly attach the string on one side with a drawing pin.

3 When the string is taut, pull it away from the wall and let go. The string will snap back smartly hitting the wall, giving a perfectly straight, coloured line.

4 Butt your stencil up to this mark and dust off the chalkline as you go. Check that the line looks correct before you start to stencil.

Disguising an Uneven Line

If you are working with a less than straight ceiling, then the method just described would highlight the unevenness. You can overcome this with the help of two friends (or one and a drawing pin). Stand back from the wall and get them to move the chalked string until a straight line is established by your eye, regardless of what the ceiling is doing. Your stencilled line will form a new decorative feature and it is this, rather than the uneven ceiling, that will be noticed.

Finding the Centre of a Square or Rectangular Panel

To find the centre of a panel, pull a chalked string diagonally over the panel, with each end of the string placed exactly in the corner (see photograph at the top of the next column). Then 'ping' the string and do the same with the opposite diagonals. An 'X' will appear where the two string marks meet and this identifies the centre of your panel. The same method applies for any square or rectangular area.

Repeating a Border Panel

If you are using a border stencil from The Stencil Library, most of them will provide this easy method of alignment. Stencils from other companies should also provide information on pattern repeats in their instruction leaflets.

1   Before removing your border stencil, mark through the registration hole with chalk or pencil (see photograph below). There is one either side of your border, so just mark the one in your direction of travel.

2   Remove the stencil and realign it by ensuring that the straight edge of your stencil is in the right position with your measured line and that the registration hole fits over the previous chalk mark (see photograph below). You are now ready to stencil your next repeat.

Negotiating Corners of Rooms

Polyester stencils will easily bend into corners. Affix the stencil to the wall with tape or repositioning adhesive. However, only stick the portion of the stencil on the wall on which you are working, and let the portion on the adjoining wall flap free. Do not try to adhere a stencil to both walls at once. Fade your colours gradually into the corner (see page 135). When complete, realign the border on the adjoining wall and let the previous portion of stencil hang free. Fade the colour in gradually from the corner and carry on as normal.

THE STENCIL LIBRARY

The Stencil Library was founded in 1988 by myself and Michael Chippendale. The company was born out of a happy combination of desire and necessity and was formulated to encompass the needs of both the amateur and the professional decorator. My partner, an interior designer who had also spent many years designing stencils for the screen printing industry, provided the technical expertise. I could not claim an art qualification other than an 'O' level, but I had enthusiasm and imagination and over the years I have developed my own design skills. 'Chips' had used stencils in our own home, but we had never conceived of the commercial possibilities. As the design climate changed in the late 1980s and the matt black culture of a few years earlier gave way to a softer, more individualistic approach to decorating, we took the plunge and launched our first catalogue of designs. It was never our aim to produce safe stencils for a mass market. Each stencil was made to order, thus negating the need to hold stock. This freed us to produce a large and diverse collection, which provided unusual subjects, styles and sizes.

Our first mail-order catalogue consisted of 250 stencils and a limited range of accessories. All the stencils were hand cut and, because the prices of the more intricate ones reflected this, we decided to offer these stencils for hire and called ourselves The Stencil Library. The catalogue was immediately well received by the home interest magazines, many of whom gave us commissions for customised designs, ensuring recognition by interior designers at an early stage. We decided to concentrate on the top end of the market and strove to design and cut to the highest quality, using the best products available. Long-standing customers will have seen our product range develop over the years.

At the time of writing, our mail-order catalogue offers a choice of approximately 1,700 stencils, distributed world-wide, with a varied range of creative supplies for the home decorator.

As our popularity increased, I found myself much in demand as a teacher. I taught part-time at a local college, combining this with public lectures, my own day courses and keeping up with the demand for our stencils. I am still an enthusiastic teacher and this book is an extension of that enjoyment. I taught myself to stencil by trial and error and had many failures as well as triumphs. Both of these results were invaluable when compiling the tips and techniques for this book.

Our continuing success meant that we had outgrown our original home. A long-held wish to move to the country saw The Stencil Library relocating to Stocksfield Hall, an eighteenth-century stone house in the tiny Northumberland settlement of Bywell, in late 1994. This became our home and the adjoining buildings our offices, school, studio and shop. My cousin Rachel also joined the company and took the top floor of the hall.

Stocksfield Hall, as well as being our home, serves a variety of other purposes. It is the blank canvas on which we test many of our designs, a source of inspiration for our students on day courses, unusual and stimulating accommodation for those on residential ones and also the venue for monthly tours and demonstrations. In fact, this book, which works its way through many of the rooms in our house, is an armchair version of one of these tours.

It took three years to reach this stage of decoration with the help of others. My professional work has always involved me in designing and decorating other people's properties, and like most of you I can only fit my own projects in during my spare time and the odd weekend.

The decoration of Stocksfield Hall is always changing. and a few of the rooms are on their second and third incarnation. As new stencil collections are designed, they are given a place in the house by painting over previous ones. My initial feelings of sadness over the demise of a room are soon overridden by the excitement of seeing the new one. Your stencilled creations might survive for many years, or you can just paint over them whenever you fancy a change. Stencilling is simply putting paint through holes in a piece of plastic or card. It helps if these holes are well designed and clean cut, but if you exercise throught, care and imagination with the basic techniques, then you have the potential to create something quite personal and unique. You may also have a great deal of fun!

The team at Stocksfield Hall: (top row) Helen, Chips; (middle row) Nikki Fionda, Lesley Thompson, Heather Phillips; (bottom row) Rachel Morris and Sabine Rose

LIST OF SUPPLIERS

The Stencil Library
Stocksfield Hall
Stocksfield
Northumberland
NE43 7TN
England
Tel: +44 1661 844 844
Fax: +44 1661 843 984
Email: sales@stencil-library.com

All stencilling requisites, varnish, paint effects and gilding supplies mentioned can be purchased from The Stencil Library in England by mail order.

Alternatively, The Stencil Library stencils and similar supplies as well as canvas floorcloth material may be purchased mail order through The Stencil Collector in the U.S. For a catalogue or further details please contact:

The Stencil Collector
1723 Tilghman Street
Allentown, PA
18104
U.S.A.
Tel & fax: 610 433 2105

Distributed in Australia through:
Stencil House Supplies Pty Ltd.
P.O. Box 141
Olinda
Victoria 3788
Tel: +61 3 9754 4040
Fax: +61 3 9754 4040

Distributed in Norway through:

Alano AS

P.O. Box 2190

N-3103 T Tønsberg

Tel: +47 33 36 29 70

Fax: +47 33 36 29 71

Email: post@alanor.no

Distributed in Germany through:

Stenbor Collections

Diening & Weigert GbR

Kirchstrabe 24

10557 Berlin

Tel: +49 30 3919542

Fax: +49 30 3919542

Email: stenbor@aol.com

Distributed in France through:

Les Trois Fontaine

57 Av. Michelet

93400 St. Ouen

France

Tel: +33 1 40 12 72 89

Fax: +33 1 40 10 24 65

Stan Pike, Blacksmith wrought

iron furniture sold through:

Arkitektura

96 Greene Street

Manhattan Island, NY

10012

Tel: 212 334 5570

Fax: 212 334 8028

Glass Paintings:

Replica

Stocksfield Hall

Stocksfield

Northumberland

NE43 7TN

England

Fax: +44 1661 843 984

Liquitex Concentrated Acrylics and Acrylic Enamels

used as stencil paint for nearest stockist:

Tel: 1–888–4ACRYLIC

INDEX